Nanoporous Carbons for Hydrogen Sorption and Electrochemical Energy Storage

Nanoporous Carbons for Hydrogen Sorption and Electrochemical Energy Storage

Guest Editors

Nikolaos Kostoglou
Claus Rebholz

MDPI

Basel • Beijing • Wuhan • Barcelona • Belgrade • Novi Sad • Cluj • Manchester

Guest Editors

Nikolaos Kostoglou
Department of Materials Science
Montanuniversität Leoben
Leoben
Austria

Claus Rebholz
Department of Mechanical and Manufacturing Engineering
University of Cyprus
Nicosia
Cyprus

Editorial Office
MDPI AG
Grosspeteranlage 5
4052 Basel, Switzerland

This is a reprint of the Special Issue, published open access by the journal *C* (ISSN 2311-5629), freely accessible at: https://www.mdpi.com/journal/carbon/special_issues/44MV18GHT2.

For citation purposes, cite each article independently as indicated on the article page online and as indicated below:

Lastname, A.A.; Lastname, B.B. Article Title. *Journal Name* **Year**, *Volume Number*, Page Range.

ISBN 978-3-7258-4663-4 (Hbk)
ISBN 978-3-7258-4664-1 (PDF)
https://doi.org/10.3390/books978-3-7258-4664-1

Contents

About the Editors vii

Preface ix

Nikolaos Kostoglou and Claus Rebhol
Editorial for Special Issue "Nanoporous Carbons for Hydrogen Sorption and Electrochemical Energy Storage
Reprinted from: *C* **2025**, *11*, 39, https://doi.org/10.3390/c11020039 1

Ravi Moreno Araujo Pinheiro Lima, Glaydson Simões dos Reis, Ulla Lassi, Eder Claudio Lima, Guilherme Luiz Dotto and Helinando Pequeno de Oliveira
Sustainable Supercapacitors Based on Polypyrrole-Doped Activated Biochar from Wood Waste Electrodes
Reprinted from: *C* **2023**, *9*, 59, https://doi.org/10.3390/c9020059 4

Milica S. Ritopečki, Natalia V. Skorodumova, Ana S. Dobrota and Igor A. Pašti
Density Functional Theory Analysis of the Impact of Boron Concentration and Surface Oxidation in Boron-Doped Graphene for Sodium and Aluminum Storage
Reprinted from: *C* **2023**, *9*, 92, https://doi.org/10.3390/c9040092 20

Ziyi Wan, Ping Tang, Luwei Dai, Yao Yang, Lu Li, Jun Liu, et al.
Highly Effective Electrochemical Water Splitting with Enhanced Electron Transfer between Ni_2Co Layered Double Hydroxide Nanosheets Dispersed on Carbon Substrate
Reprinted from: *C* **2023**, *9*, 94, https://doi.org/10.3390/c9040094 36

Nemanja Gavrilov, Stefan Breitenbach, Christoph Unterweger, Christian Fürst and Igor A. Pašti
Exploring the Impact of DAHP Impregnation on Activated Carbon Fibers for Efficient Charge Storage and Selective O_2 Reduction to Peroxide
Reprinted from: *C* **2023**, *9*, 105, https://doi.org/10.3390/c9040105 49

Marie Colin, Hani Farhat, Sam Chen, Elodie Petit, Emmanuel Flahaut, Katia Guérin and Marc Dubois
High Energy Density Primary Lithium Battery with Fluorinated S-Doped Graphene
Reprinted from: *C* **2024**, *10*, 3, https://doi.org/10.3390/c10010003 62

Florian Knabl, Nikolaos Kostoglou, Ram K. Gupta, Afshin Tarat, Steven Hinder, Mark Baker, et al.
Plasma-Treated Cobalt-Doped Nanoporous Graphene for Advanced Electrochemical Applications
Reprinted from: *C* **2024**, *10*, 31, https://doi.org/10.3390/c10020031 77

Seon Kyung Kim, Seung Ah Kim, Yoon Soo Han and Kyung-Hye Jung
Supercapacitor Performance of MXene-Coated Carbon Nanofiber Electrodes
Reprinted from: *C* **2024**, *10*, 32, https://doi.org/10.3390/c10020032 89

Rafael G. Morais, Natalia Rey-Raap, José L. Figueiredo and Manuel F. R. Pereira
Insights into the Electrocatalytic Activity of Fe,N-Glucose/Carbon Nanotube Hybrids for the Oxygen Reduction Reaction
Reprinted from: *C* **2024**, *10*, 47, https://doi.org/10.3390/c10020047 102

Kingsford Asare, Abhijeet Mali, Md Faruque Hasan, Philip Agbo, Abolghasem Shahbazi and Lifeng Zhang
Algae Derived Carbon from Hydrothermal Liquefaction as Sustainable Carbon Electrode Material for Supercapacitor
Reprinted from: *C* **2024**, *10*, 51, https://doi.org/10.3390/c10020051 . . . . **119**

About the Editors

Nikolaos Kostoglou

Nikolaos Kostoglou is a Senior Postdoctoral Researcher in the Department of Materials Science at Montanuniversität Leoben, Austria, and serves as Group Leader for Porous Materials within Prof. Christian Mitterer's research team. He received his Bachelor's and Diploma in Mechanical Engineering (2011, 2012) and an MSc in Energy Technologies and Sustainable Design with honors (2014) from the University of Cyprus, before completing a PhD with distinction (2017) at Montanuniversität Leoben, jointly supervised by Prof. Mitterer and Prof. Claus Rebholz. His research focuses on the development and advanced characterization of nanoporous carbon materials and nanocomposites for sustainable energy and environmental applications, including hydrogen storage, selective gas separation, electrochemical energy storage, water treatment and biosensing. Dr. Kostoglou has authored 49 peer-reviewed publications, participated in over 20 international conferences and contributed to multiple EU- and industry-funded projects. He has served as Guest Editor for six Special Issues and is a Topical Advisory Panel Member for *C—Journal of Carbon Research and Nanomaterials*. He has supervised numerous students and founded a research subgroup dedicated to porous materials, while building collaborations across Europe, Asia and North America. His scientific work is driven by a commitment to addressing global challenges through material innovation.

Claus Rebholz

Claus Rebholz serves as Professor of Materials & Surface Engineering and Manufacturing Processes in the Department of Mechanical and Manufacturing Engineering at the University of Cyprus. Originally from the Danube Valley in southern Germany, he earned his Diploma/BSc in Materials and Surface Engineering from the University of Applied Sciences Furtwangen in the Black Forest (1993), completing industrial placements in both Germany and Japan. He later moved to the UK on a DAAD (German Academic Exchange Service) scholarship, where he obtained an MSc (1995) and a PhD (2000) in Engineering Design and Manufacture from the University of Hull. Prof. Rebholz spent six years with the Robert Bosch Group. He began as a Research & Development Executive Management Trainee, working at research centers in Germany, Japan and the US (1998–2001). He later led a multidisciplinary profit center in laser materials processing in Stuttgart/Germany (2001–2004), before joining the faculty at the University of Cyprus. His research focuses on the processing–structure–properties relationships in tailor-made functional materials for automotive, energy, environmental and nanomanufacturing applications. Much of this work has been conducted in collaboration with industry and international research groups. Among his recognitions, Prof. Rebholz received the European Commission Marie Curie Excellence Team Leader Award for his work on nano-heater systems for thermal nanomanufacturing. He also holds Visiting Academic/Professor positions in Materials Science at Montanuniversität Leoben (Austria), in Materials Chemistry at RWTH Aachen (Germany) and in Materials at the University of Manchester (UK).

Preface

The Reprint, titled *"Nanoporous Carbons for Hydrogen Sorption and Electrochemical Energy Storage"*, brings together a variety of pioneering studies at the intersection of carbon science and sustainable energy technologies. This body of work reflects the growing significance of nanoporous carbon materials in addressing critical challenges in hydrogen storage, electrocatalysis and electrochemical energy storage. The Reprint's scope spans experimental investigations and computational modeling of advanced carbon structures designed to enhance charge storage capacity, catalytic activity and energy efficiency. It is aimed at scientists, engineers and graduate students who are interested in the development of next-generation materials for clean energy applications. The motivation behind compiling this Reprint was to provide a focused platform showcasing recent progress and inspiring future innovations in the design and functionalization of carbon-based materials. We hope it will serve as a comprehensive reference and catalyst for continued research in the field.

Nikolaos Kostoglou and Claus Rebholz
Guest Editors

C Journal of Carbon Research

MDPI

Editorial

Editorial for Special Issue "Nanoporous Carbons for Hydrogen Sorption and Electrochemical Energy Storage

Nikolaos Kostoglou [1,*] and Claus Rebholz [2]

1 Department of Materials Science, Montanuniversität Leoben, 8700 Leoben, Austria
2 Department of Mechanical and Manufacturing Engineering, University of Cyprus, 1678 Nicosia, Cyprus; rebholz.claus@ucy.ac.cy
* Correspondence: nikolaos.kostoglou@unileoben.ac.at

Increasing global energy demand and the need for sustainable energy solutions have sparked significant interest in carbon-based materials for energy conversion and storage [1,2], along with hydrogen-related technologies [3,4]. Among these, nanoporous carbons stand out due to their tunable pore structure, high surface area, and ease of surface functionalization [5–7]. These unique properties make them excellent candidates for enhancing the performance of electrochemical systems and electrocatalytic devices by improving charge transport, ion accessibility, and catalytic activity [8–11]. This Special Issue of *C—Journal of Carbon Research* gathers original contributions exploring both fundamental and applied aspects of carbon-based materials in electrocatalysis—specifically for water splitting and oxygen reduction—as well as electrochemical energy storage using supercapacitors and batteries. The featured articles address diverse research challenges and innovations in the field of nanoporous carbon-based materials for energy-relevant reactions, including the oxygen reduction reaction (ORR), oxygen evolution reaction (OER) and hydrogen evolution reaction (HER), alongside breakthroughs in battery and supercapacitor technologies. The articles are as follows:

- Morais et al. [12] developed Fe,N-doped, glucose-derived carbon nanotube hybrids, demonstrating enhanced electrocatalytic activity for the ORR through synergetic incorporation of Fe and N.
- Asare et al. [13] introduced algae-derived carbon obtained via hydrothermal liquefaction as a sustainable electrode material, highlighting its excellent long-term supercapacitor performance following KOH activation.
- Lima et al. [14] produced polypyrrole-doped activated biochar from wood waste, offering a sustainable pathway to high-performance supercapacitor electrodes.
- Ritopecki et al. [15] applied Density Functional Theory to study the impact of B concentration and surface oxidation in B-doped graphene, revealing significant improvements in Na and Al ion storage capacities.
- Wan et al. [16] reported the fabrication of Ni_2Co layered double hydroxide nanosheets on carbon substrates for water splitting, showcasing low overpotentials and high performance for both HER and OER through enhanced electron transfer.
- Gavrilov et al. [17] explored diammonium hydrogen phosphate-impregnated activated carbon fibers, achieving selective oxygen reduction to hydrogen peroxide while optimizing capacitive behavior.
- Colin et al. [18] presented fluorinated S-doped graphene as a cathode material for high-energy lithium batteries, attaining outstanding power and energy densities.

- Knabl et al. [19] synthesized Co-doped nanoporous graphene via plasma treatment, achieving multifunctionality for both OER-driven water splitting and supercapacitor applications.
- Kim et al. [20] demonstrated the promising performance of MXene-coated carbon nanofibers as pseudocapacitive electrodes, combining high specific capacitance with excellent stability.

Together, these articles provide a comprehensive overview of current advances in the synthesis, functionalization, and application of nanoporous carbon-based materials. They highlight the crucial role of material design and interface engineering in tuning electrochemical and catalytic properties for real-world energy applications.

We sincerely thank all contributing authors and the reviewers for their excellent efforts and insights. We also extend our appreciation to the editorial team of C for their continuous support. We hope this Special Issue will serve as a valuable reference for researchers in the fields of carbon science, hydrogen technologies, and electrochemical energy storage.

Conflicts of Interest: The editors declare no conflicts of interest.

References

1. Kothandam, G.; Singh, G.; Guan, X.; Lee, J.M.; Ramadass, K.; Joseph, S.; Benzigar, M.; Karakoti, A.; Yi, J.; Kumar, P.; et al. Recent advances in carbon-based electrodes for energy storage and conversion. *Adv. Sci.* **2023**, *10*, 2301045. [CrossRef] [PubMed]
2. Xiao, J.; Han, J.; Zhang, C.; Ling, G.; Kang, F.; Yang, Q.H. Dimensionality, function and performance of carbon materials in energy storage devices. *Adv. Energy Mater.* **2022**, *12*, 2100775. [CrossRef]
3. Mohan, M.; Sharma, V.K.; Kumar, E.A.; Gayathri, V.J.E.S. Hydrogen storage in carbon materials—A review. *Energy Storage* **2019**, *1*, e35. [CrossRef]
4. Zhou, W.; Jia, J.; Lu, J.; Yang, L.; Hou, D.; Li, G.; Chen, S. Recent developments of carbon-based electrocatalysts for hydrogen evolution reaction. *Nano Energy* **2016**, *28*, 29–43. [CrossRef]
5. Kostoglou, N.; Koczwara, C.; Stock, S.; Tampaxis, C.; Charalambopoulou, G.; Steriotis, T.; Paris, O.; Rebholz, C.; Mitterer, C. Nanoporous polymer-derived activated carbon for hydrogen adsorption and electrochemical energy storage. *Chem. Eng. J.* **2022**, *427*, 131730. [CrossRef]
6. Gadipelli, S.; Howard, C.A.; Guo, J.; Skipper, N.T.; Zhang, H.; Shearing, P.R.; Brett, D.J. Superior multifunctional activity of nanoporous carbons with widely tunable porosity: Enhanced storage capacities for carbon-dioxide, hydrogen, water, and electric charge. *Adv. Energy Mater.* **2020**, *10*, 1903649. [CrossRef]
7. Benzigar, M.R.; Talapaneni, S.N.; Joseph, S.; Ramadass, K.; Singh, G.; Scaranto, J.; Ravon, U.; Al-Bahily, K.; Vinu, A. Recent advances in functionalized micro and mesoporous carbon materials: Synthesis and applications. *Chem. Soc. Rev.* **2018**, *47*, 2680–2721. [CrossRef] [PubMed]
8. Shao, H.; Wu, Y.C.; Lin, Z.; Taberna, P.L.; Simon, P. Nanoporous carbon for electrochemical capacitive energy storage. *Chem. Soc. Rev.* **2020**, *49*, 3005–3039. [CrossRef] [PubMed]
9. Tan, H.; Tang, J.; Kim, J.; Kaneti, Y.V.; Kang, Y.M.; Sugahara, Y.; Yamauchi, Y. Rational design and construction of nanoporous iron-and nitrogen-doped carbon electrocatalysts for oxygen reduction reaction. *J. Mater. Chem. A* **2019**, *7*, 1380–1393. [CrossRef]
10. Sathiskumar, C.; Ramakrishnan, S.; Vinothkannan, M.; Rhan Kim, A.; Karthikeyan, S.; Yoo, D.J. Nitrogen-doped porous carbon derived from biomass used as trifunctional electrocatalyst toward oxygen reduction, oxygen evolution and hydrogen evolution reactions. *Nanomaterials* **2019**, *10*, 76. [CrossRef] [PubMed]
11. Young, C.; Lin, J.; Wang, J.; Ding, B.; Zhang, X.; Alshehri, S.M.; Ahamd, T.; Salunkhe, R.R.; Hossain, S.A.; Khan, J.H.; et al. Significant effect of pore sizes on energy storage in nanoporous carbon supercapacitors. *Chem.–A Eur. J.* **2018**, *24*, 6127–6132. [CrossRef] [PubMed]
12. Morais, R.G.; Rey-Raap, N.; Figueiredo, J.L.; Pereira, M.F.R. Insights into the Electrocatalytic Activity of Fe,N-Glucose/Carbon Nanotube Hybrids for the Oxygen Reduction Reaction. *C* **2024**, *10*, 47. [CrossRef]
13. Asare, K.; Mali, A.; Hasan, M.F.; Agbo, P.; Shahbazi, A.; Zhang, L. Algae Derived Carbon from Hydrothermal Liquefaction as Sustainable Carbon Electrode Material for Supercapacitor. *C* **2024**, *10*, 51. [CrossRef]
14. Lima, R.M.A.P.; dos Reis, G.S.; Lassi, U.; Lima, E.C.; Dotto, G.L.; de Oliveira, H.P. Sustainable Supercapacitors Based on Polypyrrole-Doped Activated Biochar from Wood Waste Electrodes. *C* **2023**, *9*, 59. [CrossRef]
15. Ritopecki, M.S.; Skorodumova, N.V.; Dobrota, A.S.; Pašti, I.A. Density Functional Theory Analysis of the Impact of Boron Concentration and Surface Oxidation in Boron-Doped Graphene for Sodium and Aluminum Storage. *C* **2023**, *9*, 92. [CrossRef]

16. Wan, Z.; Tang, P.; Dai, L.; Yang, Y.; Li, L.; Liu, J.; Yang, M.; Deng, G. Highly Effective Electrochemical Water Splitting with Enhanced Electron Transfer between Ni2Co Layered Double Hydroxide Nanosheets Dispersed on Carbon Substrate. *C* **2023**, *9*, 94. [CrossRef]
17. Gavrilov, N.; Breitenbach, S.; Unterweger, C.; Fürst, C.; Pašti, I.A. Exploring the Impact of DAHP Impregnation on Activated Carbon Fibers for Efficient Charge Storage and Selective O_2 Reduction to Peroxide. *C* **2023**, *9*, 105. [CrossRef]
18. Colin, M.; Farhat, H.; Chen, S.; Petit, E.; Flahaut, E.; Guérin, K.; Dubois, M. High Energy Density Primary Lithium Battery with Fluorinated S-Doped Graphene. *C* **2024**, *10*, 3. [CrossRef]
19. Knabl, F.; Kostoglou, N.; Gupta, R.K.; Tarat, A.; Hinder, S.; Baker, M.; Rebholz, C.; Mitterer, C. Plasma-Treated Cobalt-Doped Nanoporous Graphene for Advanced Electrochemical Applications. *C* **2024**, *10*, 31. [CrossRef]
20. Kim, S.K.; Kim, S.A.; Han, Y.S.; Jung, K.-H. Supercapacitor Performance of MXene-Coated Carbon Nanofiber Electrodes. *C* **2024**, *10*, 32. [CrossRef]

C Journal of Carbon Research

Article

Sustainable Supercapacitors Based on Polypyrrole-Doped Activated Biochar from Wood Waste Electrodes

Ravi Moreno Araujo Pinheiro Lima [1,†], Glaydson Simões dos Reis [2,*,†], Ulla Lassi [3,4,*], Eder Claudio Lima [5], Guilherme Luiz Dotto [6] and Helinando Pequeno de Oliveira [1,*]

1 Institute of Materials Science, Federal University of Sao Francisco Valley, Juazeiro 48902-300, Bahia, Brazil; raviplima.engmec@gmail.com
2 Department of Forest Biomaterials and Technology, Swedish University of Agricultural Sciences, Biomass Technology Centre, SE-90183 Umeå, Sweden
3 Research Unit of Sustainable Chemistry, University of Oulu, FI-90014 Oulu, Finland
4 Unit of Applied Chemistry, University of Jyvaskyla, Kokkola University Consortium Chydenius, Talonpojankatu 2B, FI-67100 Kokkola, Finland
5 Institute of Chemistry, Federal University of Rio Grande do Sul–UFRGS, Av. Bento Gonçalves 9500, P.O. Box 15003, Porto Alegre 91501-970, RS, Brazil; profederlima@gmail.com
6 Research Group on Adsorptive and Catalytic Process Engineering (ENGEPAC), Federal University of Santa Maria, Av. Roraima, 1000-7, Santa Maria 97105–900, RS, Brazil; guilherme_dotto@yahoo.com.br
* Correspondence: glaydson.simoes.dos.reis@slu.se (G.S.d.R); ulla.lassi@oulu.fi (U.L.); helinando.oliveira@univasf.edu.br (H.P.d.O.)
† These authors contributed equally to this work.

Abstract: The synthesis of high-performance carbon-based materials from biomass residues for electrodes has been considered a challenge to achieve in supercapacitor-based production. In this work, activated biochar has been prepared as the active electrode material for supercapacitors (SCs), and an effective method has been explored to boost its capacitive performance by employing polypyrrole (PPy) as a biochar dopant. The results for physicochemical characterization data have demonstrated that PPy doping affects the biochar morphology, specific surface area, pore structure, and incorporation of surface functionalities on modified biochar. Biochar-PPy exhibited a surface area of 87 $m^2\ g^{-1}$, while pristine biochar exhibited 1052 $m^2\ g^{-1}$. The SCs were assembled employing two electrodes sandwiched with PVA solid-state film electrolyte as a separator. The device was characterized by standard electrochemical assays that indicated an improvement of 34% in areal capacitance. The wood electrodes delivered high areal capacitances of 282 and 370 mF cm^{-2} at 5 mA cm^{-2}, for pure biochar and biochar doped with PPy, respectively, with typical retention in the capacitive response of 72% at the end of 1000 cycles of operation of the supercapacitor at high current density, indicating that biochar-PPy-based electrode devices exhibited a higher energy density when compared to pure biochar devices.

Keywords: wood waste; wood electrodes; wood-based supercapacitors; polypyrrole; pseudocapacitance

1. Introduction

The growing global population, the depletion of natural resources, and the negative impact of industrial manufacturing processes are driving factors for the increasing demand for more sustainable manufacturing processes and products for the energy storage industry. As a consequence, scientists have explored solutions for sustainable materials and products for greener storage devices through a tenfold increase in scientific publications in 2010–2020 [1–5].

Due to the diversity of the required small and large energy systems, the development of electrochemical energy storage devices, such as supercapacitors (SCs) and batteries, has been extensively reported in the literature; batteries have higher energy densities at the cost of low power densities, while SCs are characterized by high power densities and low energy densities, with the advantages of having a high power capacity and being endowed with fast charging/discharging cycles [6], light weight [5], and both economic and environmental advantages such as the use of aqueous electrolytes [2]. SCs have a huge relevance for applications such as electrical automobiles [7] and telecommunications [8] by integrating strategies based on the incorporation of solar electricity or the harvesting of mechanical movement converted into electricity [5,9,10].

The optimal configuration of supercapacitors has been explored with the combination of porous electrodes and electrolytes to improve the charge separation at the Helmholtz double layer and the electrode–electrolyte interface [4]. Supercapacitors are composed of electrodes that depend on features such as high specific surface area, surface chemistry, and electrical conductivity [11]. To date, several materials (e.g., graphite [12], metal oxides/hydroxides [13], and conducting polymers [14]) have been employed as high-performance electrodes for SCs. On the other hand, two-dimensional nanostructures (graphene, MXene, metal dichalcogenides) present outstanding properties such as high package density, high surface area, transparency, and chemical/mechanical stability that enable their use in supercapacitors [15]. Despite these advantages, the typical stacking/aggregation of 2D structures reduces the electrochemical performance of the devices. To circumvent these drawbacks, different strategies have focused on the exfoliation of black phosphorous [16] and on the surface modification of structures with nitrogen, sulfur, and phosphorous-based groups (heteroatom doping strategies) that reinforce the pseudocapacitance of the resulting material [17]. The incorporation of materials with characteristic redox reactions is another important strategy that is conducted given the fast electrochemical kinetics in systems such as vanadium redox flow batteries [18] and with the incorporation of carbonaceous materials to avoid corrosive and degradative processes in Zn anodes [19].

However, these electrode materials have the drawbacks of high production costs and/or non-eco-friendly fabrication methods. Thus, it is crucial to develop electrode materials with the properties of sustainability, eco-friendly behavior, low cost, and efficient response. In this direction, carbon electrodes from biomass precursors have attracted huge attention due to their worldwide availability and abundance, non-toxicity, and high surface area with hierarchical porous structure materials [20,21] to improve the mechanisms of efficient charge separation.

Pyrolysis is a standard method to convert biomass into porous materials whose properties are highly dependent on the pyrolysis conditions and the chemicals used for the activation or doping processes [22–24]. For instance, nitrogen doping methods in carbon preparation for SCs have been shown to boost its conductivities and hydrophilicity, which improve its surface wettability, resulting in an increased pseudocapacitance effect, thus delivering improved energy/power densities.

One of the ways to improve the electrochemical conductivity of carbon-based electrodes is coating with conducting polymers [25–27], such as polypyrrole (PPy), which is one of the most promising support materials due to its excellent electrical conductivity, environmental and thermal stability, and easy procedure of preparation [28–32]. Thus, it enables a promising possibility for the development of low-cost and commercially viable SCs.

The general mechanisms that are combined in the improvement of the electrochemical efficiency of the electrodes are based on two important processes: the electrical double layer capacitance (EDLC) acquired from the adsorption of electrolyte ions on conductive electrodes [33] and pseudocapacitive effects that allow good performance in energy storage through reversible redox on electrodes [34]. The adequate combination of EDLC and the pseudocapacitive effect in SCs improves the overall energy density of the devices by the combination of available sites for charge accumulation of EDLC and the rapid ionic transport of pseudocapacitive prototypes provided by the PPy-based biochar doping [35–37].

Herein, we aimed to explore the use of birch wood to produce porous biochar to be used as carbon electrodes for high-performance supercapacitors. Moreover, the prepared porous biochar was subjected to a polypyrrole, and its effect on both physicochemical and electrochemical properties was fully investigated. The obtained results suggest an improved efficiency of the polypyrrole-doped biochar due to the pseudo-capacitive effect and the feasibility of the proposed approach for the fabrication of sustainable SCs based on biomass wastes.

2. Materials and Methods

2.1. Materials

Polyvinylidene fluoride (PVDF), cetyl trimethyl ammonium bromide (CTAB), dimethylformamide (DMF), and pyrrole (PPy) were obtained from Sigma Aldrich (St. Louis, MO, USA). Ammonium persulfate (APS) was obtained from Química Moderna (Barueri, SP, Brazil). Carbon black was purchased from Micromeritics (Norcross, GA, USA). All materials were used as received, except for the pyrrole, which was distilled before use.

2.2. Biochar Preparation

The biochar was prepared using birch wood wastes as a precursor. First, 20.0 g of the dried biomass was mixed with H_3PO_4 (50%) at a ratio of 1:4 (weight) and mixed until forming a homogeneous paste [38,39]. Then, the paste was kept at room temperature for 2 h and dried at 105 °C overnight. The dried paste was pyrolyzed at 700 °C for 2 h under an N_2 atmosphere, with an initial heating rate of 10 °C per min. The pyrolyzed material was ground and washed several times with boiling water until the pH value of the filtrate water was similar to the ultra-pure water.

2.3. Preparation of Birch-PPy Powder

The birch-PPy powder was prepared according to the procedure described as follows: birch powder was dispersed into 5 mL of Milli-Q water by sonication. Meanwhile, another solution containing polypyrrole and CTAB was prepared as follows: 35 uL of pyrrole was added into 5 mL of Milli-Q water and also incorporated with 9.6 mg of CTAB. This solution was stirred until the complete dispersion of the CTAB and mixed into the solution containing birch biochar to prepare solution A. After this, solution B was prepared by adding 114.6 mg of APS into 5 mL of milli-Q water. Then, solution A was stirred in an ice bath and received solution B (dropwise) in a process that initiated the polymerization. After 2 h of reaction, the dark solution was centrifuged at 5000× g for 5 min to obtain the precipitated particles. The black powder was filtered and washed several times with Milli-Q water to remove any residue from the polymerization process and was dried in an oven at 40 °C for 1 h. Finally, the birch-PPy at ambient temperature was stored for use. The loaded mass of polypyrrole in the composite was 11 mg.

2.4. Fabrication of the Biochar SCs Electrodes

Graphite paper (1 × 1 cm) was used as a support for the coating with the biochar-based material slurries, prepared with a mass ratio of 8:1:1 (biochar:PVDF:carbon black) as follows: 10 mg of PVDF was added to 500 μL of DMF and then heated to 60 °C and stirred until the PVDF was completely dispersed. After this step, 80 mg of birch biochar and 10 mg of carbon black were added to the mixture, which was kept under continuous stirring at 60 °C to provide a homogeneous dispersion of the carbonaceous derivative. The preparation of the electrode, after this step, was conducted as follows: First, 20 μL of the as-prepared slurry was dropped and spread on the substrate. Then, the coated graphite paper was heated at 50 °C to eliminate the residues of the solvent. The biochar-PPy-based electrodes were prepared using the same method, although instead of using the pure birch biochar, the composite biochar-PPy was used for the preparation of the slurry.

2.5. Electrolyte Preparation and Supercapacitor Assembly

Polyvinyl alcohol (PVA) was explored as a solid-state film for the preparation of the electrolyte. The standard procedure for preparation was conducted as follows: First, 1 g of PVA was added into 10 mL of Milli-Q water and heated to 70 °C under stirring for 1 h. After this, the temperature was raised to 100 °C, and the solution was stirred for an additional 1 h. Then, a transparent aspect of the solution was observed, and the solution was cooled to room temperature. As a following step, the solution was stirred for 15 min and received 1 mL of H_3PO_4. To avoid bubbles, the solution was sonicated for 2 min. Then, the solution was placed in a mold, and a thin film of PVA was obtained after 48 h. The PVA film was cut into pieces with 1 cm^2 of area and used as a solid-state electrolyte. The SCs were assembled in a sandwich configuration in which two electrodes were placed parallel to each other, separated by the PVA solid electrolyte film.

2.6. Characterization

The morphology evaluation of the biochars was carried out using a scanning electron microscope Vega 3XM (Tescan) with an electron acceleration of 10 kV, with the collected images collected with magnifications of 500 x, 1 kx, 3 kx, and 5 kx. The chemical composition of the materials was evaluated from Fourier transform infrared spectrum (FTIR) using the KBr method in an IR Prestige-21 Fourier transform infrared spectrometer (Shimadzu).

The specific surface area and pore volume of the biochar materials were measured through N_2 adsorption-desorption via the BET (Brunauer, Emmett and Teller) and BJH (Barrett-JoynerHalenda) method, respectively, on a Surface Area Analyzer (ASAP 2020, Micromeritics).

Electrochemical characterization of the supercapacitors explored the two-electrode configuration with measurements provided by a potentiostat Autolab PGSTAT 302 N (Methrom) for the acquisition of voltammetry curves at different scan rates (10 mV s^{-1} to 200 mV s^{-1}), with a potential window range of 0 to 0.8 V. Galvanometric curves was performed with a current density ranging from 1 mA to 5 mA, and the impedance spectrum was evaluated in a frequency range of 1 Hz to 1 MHz with data fitted by modified Randles circuit using the software Zview 2 version 3.4.

The areal capacitance was calculated from galvanostatic curves, as follows:

$$C_a = \frac{2A_{discharge} \times I}{V^2 \times A_{electrode}} \quad (1)$$

where $A_{discharge}$ is the area of the discharge curve, I is the current applied, V is the potential relative to the beginning of the discharge curve (below the IR drop), and $A_{electrode}$ is the area of the electrode.

To obtain the Ragone data, the calculus of the energy and power density was provided by Equations (2) and (3):

$$E_d = \frac{C_a \times V^2}{2 \times 3600} \quad (2)$$

$$P_d = \frac{3600 \times E_d}{\Delta t} \quad (3)$$

where E_d is the energy density (W h cm^{-2}), P_d is the power density (W cm^{-2}), C_a is the areal capacitance obtained from Equation (1), and Δt is the discharge time.

3. Results and Discussion

3.1. Characterization of the Biochar and Biochar-PPy Electrode Materials

The surface morphology of the biochar and biochar doped with PPy was evaluated from SEM analyses. The SEM images at different magnifications are shown in Figure 1. The pristine biochar (Figure 1a) is characterized by a dense distribution of particles with a low concentration of small grains on its surface, while biochar doped with polypyrrole (see Figure 1b) shows an evident change in morphology due to the coating of biochar by aggregates of grains of polypyrrole (a typical disposition of PPy grains), showing that the doping process provoked important changes in the biochar morphology. Such difference is evidence of the adherence of polypyrrole on biochar's surface, which could create both defects in biochar structure and nitrogen functionalities (from PPy) that can boost its electrochemical performances [40]. Overlaid EDS images in Figure 1c,d for biochar and biochar-PPy, respectively, confirm the higher density of nitrogen-based groups by coating with polypyrrole (green dots are attributed to carbon and red dots to nitrogen), confirming the adequate modification with PPy.

Figure 1. SEM images for birch biochar (**a**) and biochar doped with PPy (**b**), and overlaid EDS images (red dots for nitrogen element and green dots for carbon element) for biochar (**c**) and biochar doped with PPy (**d**).

Images for as-prepared electrodes of biochar (Figure 2a) and biochar doped with PPy (Figure 2b) correspond to the images reported for composites (Figure 1) in which a dense distribution of grains is observed for the coated material (incorporation of PPyon biochar). After 1000 cycles of use, the electrodes were re-evaluated (images shown in Figure 2c (biochar) and Figure 2d (biochar +PPy)), indicating a more compact structure with a reduction in the size of grains of biochar and compressed grains of polypyrrole, as a result of the compressive forces during measurement and the smoothness of polypyrrole. No cracks of microstructural changes were observed as a result of the electrochemical characterization.

Figure 2. Comparison between SEM images for electrodes before use of birch biochar (**a**) and biochar doped with PPy (**b**), and after 1000 cycles of operation—birch biochar (**c**) and biochar doped with PPy (**d**).

The pore structure and the change in the porosity degree before and after PPy doping were evaluated following the N_2 adsorption/desorption as shown in Figure 3. The N_2 adsorption-desorption isotherms of biochar (Figure 3a) and biochar-PPy (Figure 3b) show that the PPy doping provoked a high impact on their shapes and the amount of adsorbed N_2. According to the IUPAC classification [41], the non-doped biochar shows an isotherm that seems to be a mix of types I and IV; type I is related to the presence of micropores due to the high N_2 adsorption at low partial pressure, and type IV since it shows hysteresis from partial relative pressure of 0.4, which is a strong indication of the presence of mesoporosity [23,42]. The biochar doped with PPy shows a clear isotherm type IV [41]. Moreover, the deposition of PPy on biochar structure seems to reduce the porosity of the biochar-PPy due to reduced adsorbed N_2 volume: 458.0 $cm^3\ g^{-1}$ and 57.0 $cm^3\ g^{-1}$ for biochar and biochar-PPy, respectively.

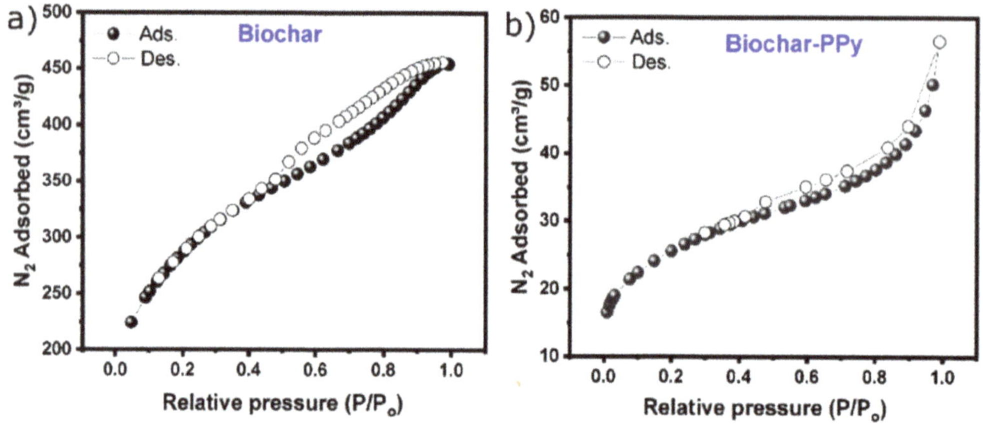

Figure 3. Nitrogen (N_2) adsorption/desorption isotherms of (**a**) biochar and (**b**) Biochar-PPy.

The SSA and porosity values of the biochar electrodes are shown in Table 1. Biochar-PPy had a surface area of 87 $m^2\ g^{-1}$, which is much lower than that of pristine biochar (1052 $m^2\ g^{-1}$) (see Table 1). The decrease in the SSA and pore volume was due to the deposition of PPy on the carbon structure that covered the highly porous surface of the biochar sample.

Table 1. Textural properties of the biochar samples.

Samples	SSA ($m^2\ g^{-1}$)	A_{Micro} ($m^2\ g^{-1}$)	A_{Meso} ($m^2\ g^{-1}$)	Pore Volume ($cm^3\ g^{-1}$)
Biochar	1052	240	812	0.46
Biochar-PPy	87	21	66	0.072

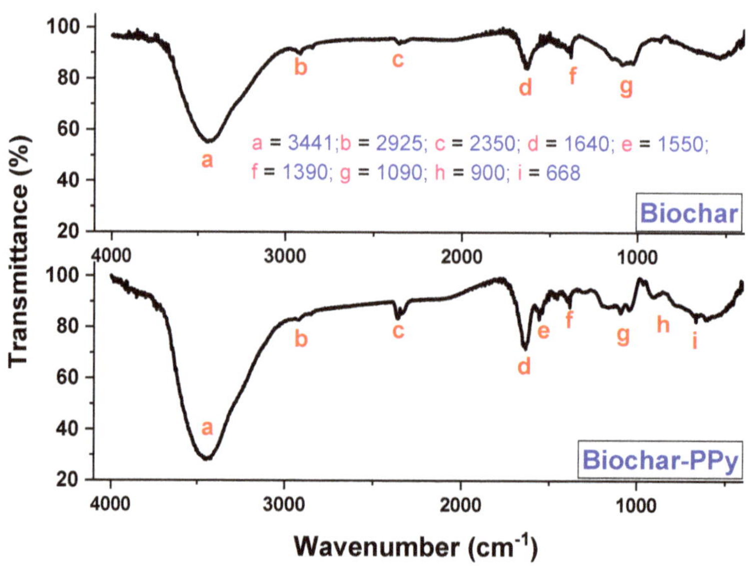

Figure 4. FTIR spectra of biochar and biochar-PPy samples.

The reduction of SSA can be beneficial for SC performance since the pseudocapacitance of polypyrrole prevails on the EDLC performance of carbon derivatives, which is surface area dependent. As expected, the insertion of PPy and N-functionalities can boost the pseudocapacitance of the biochar-PPy through redox reactions.

FTIR spectra were evaluated to examine the presence of surface functionalities on biochar surfaces, as well as the impact of PPy doping on the resulting material. FTIR spectra of biochar and biochar modified with PPy are shown in Figure 4. It is observed that the PPy doping caused an impact on biochar's surface functionalities. The absorption bands at 3441 cm^{-1} may be attributed to the O−H stretching vibration and are broader in biochar-PPy. The small peak at 2925 cm^{-1} is related to symmetric vibrations of the CH_2 units [43]. The peak at 1640 cm^{-1} is due to the stretching vibrations of the C=O and C=C units [44]. The spectrum of the biochar-PPy exhibited characteristic peaks at 1550, 900, and 668 cm^{-1}, which are indexed to the C=C stretching, C–N stretching, and C–N in-plane deformation, respectively, indicating the presence of a doped PPy state [45]. The above outcomes suggest that the biochar-PPy material was successfully prepared.

3.2. Electrochemical Assays

The electrochemical performance of supercapacitors can be successfully evaluated by cyclic voltammogram (CV) and galvanostatic charge/discharge (GCD) analyses. CV profiles of the biochar and biochar doped with PPy were evaluated using the electrochemical cell in which supercapacitor prototypes are sandwiched in parallel, plated, and characterized at scan rates ranging from 10 to 200 mV s^{-1} (see Figure 5a,b). Both curves show that square-shaped aspects are more pronounced at a lower scan rate. At increasing scan rates, the oblate aspect of the curves can be assigned to the pseudocapacitive behavior of additives. In terms of the maximum value of the current at a corresponding value of the scan rate, it is possible to observe that values for polypyrrole-based supercapacitors are higher than those observed for non-modified devices (biochar-based supercapacitors). As a consequence, the enclosed area in curves (applied as a parameter for the calculus of the areal capacitance) indicates that the incorporation of polypyrrole introduces changes in the electrochemical performance of the modified electrodes. These cone-shaped curves, observed at a high scan rate (200 mVs^{-1}), are attributed to the abundance of functional groups that are known to be present on biochars, especially biochars doped with PPy [44]. As expected, the incorporation of PPy in the structure contributes to reversible redox reactions that favor an increase in the overall capacitance and consequently in the energy density due to the facilitated transference of ions under the doping process in cycles of swelling/contraction of the conducting polymer chain.

The electrochemical performance of electrodes was further evaluated by using GCD curves (see Figure 5c,d). Both biochar and biochar-PPy showed slight deviations from the linear branch. For PPy-doped electrodes, a slight curvature is observed, indicating the reversible accumulation of charge and a pseudocapacitance effect due to the redox reaction provoked by the nitrogen functionalities on biochar-PPy. As expected, a complete charge-discharge takes longer characteristic times for supercapacitors under a lower current density regimen. The supercapacitor based on biochar-PPy displayed a total charge-discharge period in the order of 680 s, while pure biochar is in the order of 510 s (at a corresponding current of 1 mA), highlighting the importance of the polypyrrole doping on the overall SCs' performance. These results following CV analysis indicated a superior electrochemical performance for biochar doped with PPy.

The areal capacitances obtained from the GCD curves and calculated using Equation (1) (and shown in Figure 6a) confirm that for all of the current density ranges, the performance for PPy-modified biochar-based electrodes is better than that observed for pristine ones. The best performance in the overall range of current densities is the result of the combined good pseudocapacitive contribution and the high conductivity of the polymeric film. An interesting aspect to be considered from these results is that a slight reduction in the performance of biochar-based supercapacitors at increasing current density is circumvented by a slight increase in the performance of PPy-modified biochars, which is clear evidence that operation at higher current density regimen is favored by the incorporation of a conducting polymer on devices. The corresponding variation for areal capacitance as a function of the scan rate is shown in Figure 6b. As can be seen, an overall reduction in the performance is observed in both experimental systems at increasing scan rate [37], with a general best performance for biochar-PPy samples, confirming the previous results from GCD data.

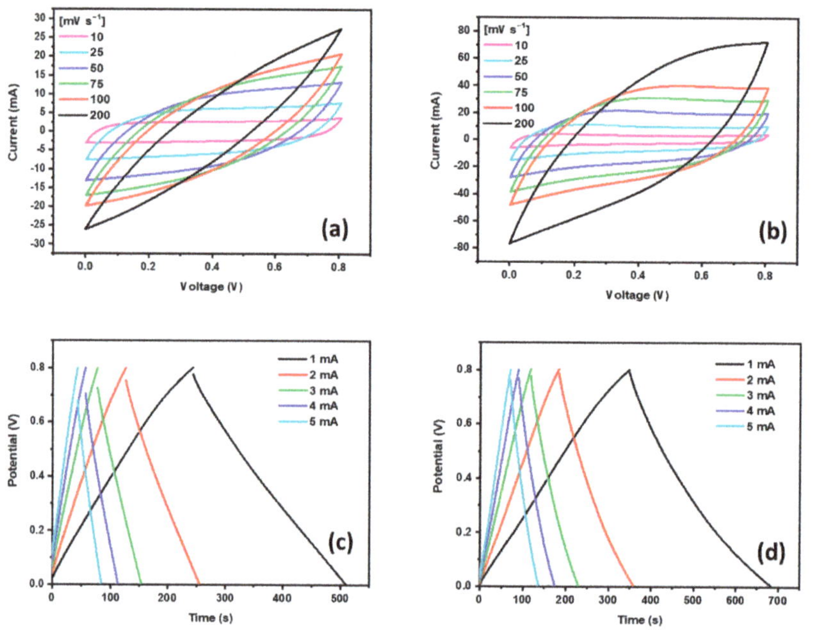

Figure 5. CV curves at different scan rates for biochar (**a**) and biochar-PPy-based electrodes (**b**), and GCD curves at various current densities for biochar (**c**) and biochar-PPy (**d**) electrodes.

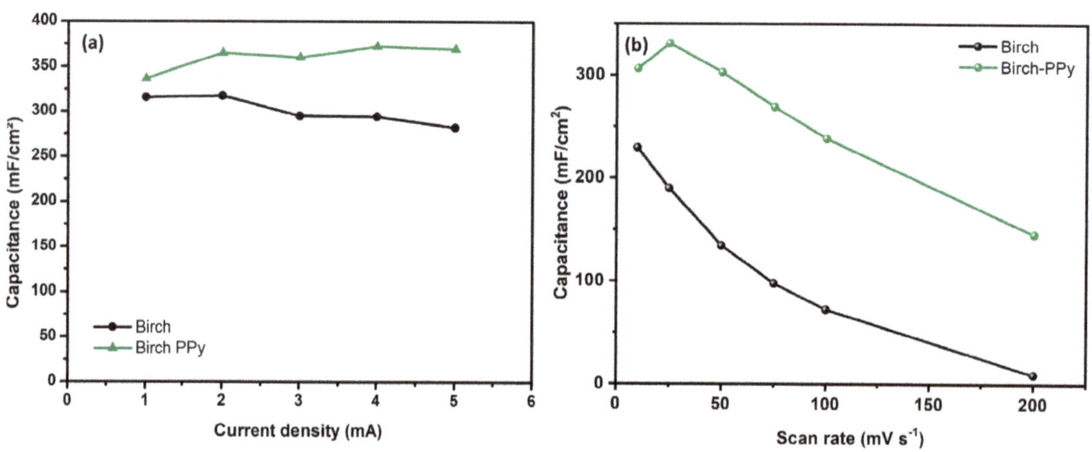

Figure 6. Areal capacitances at different current densities for supercapacitors based on biochar and biochar-PPy electrodes calculated from GCD curves (**a**) and corresponding variation of the areal capacitance as a function of the scan rate in CV curves (**b**).

The performance of modified supercapacitors introduces advantages in comparison with corresponding systems reported in the literature, such as polyaniline-doped graphene fiber [46] and $NiCo_2O_4$ thin film [47]. These materials presented complex synthesis routes and still presented poorer electrochemical properties in comparison with the experimental system reported in this work. A more complete comparison with different electrode materials is given in Table 2, which provides a comparison of the C_A values of several types of electrode materials. As can be seen, a good performance for non-doped and PPy-doped biochar electrodes is observed by direct comparison with supercapacitors prepared with different electrodes. These highlights indicate that birch wastes can be a low-cost, sustainable, and efficient precursor to fabricate high-performance electrodes for SCs application.

Table 2. The areal capacitance of different electrode materials in comparison with that of the present work.

Electrode Material	Areal Capacitance	Capacitance Retention/Cycle	Electrolyte	Current Density ($A\ g^{-1}$) or Scan Rate ($mV\ s^{-1}$)	Ref.
Graphene fiber	3.3 mF cm^{-2}	-/5000	PVA/H_3PO_4	0.1 mA cm^{-2}	[46]
Polyaniline-doped graphene fiber	66.6 mF cm^{-2}	-/5000	PVA/H_3PO_4	0.1 mA cm^{-2}	[46]
$NiCo_2O_4$ thin film	40.6 mF cm^{-2}	96.5%/10,000	2 M KOH	0.133 mA cm^{-2}	[47]
MnO_2 /MoS_2	224 mF cm^{-2}	90%/3000	1.0 M Na_2SO_4	0.1 mA cm^{-2}	[48]
TiO_2	23.24 mF cm^{-2}	-/10,000	0.5 M Na_2SO_4	2 mV s^{-1}	[49]
Sheet-like $ZnCo_2O_4$	16.13	-/1000	1 M KOH	0.01 mA cm^{-2}	[50]
Carbon dots/graphene microfibers	607 mF cm^{-2}	-/2000	$EMIBF_4$/PVDF-HFP	20 mA cm^{-2}	[51]
MnO_2-modified hierarchical graphene fiber	9.6 mF cm^{-2}	-/1000	H_2SO_4–PVA	10 mV s^{-1}	[52]
Graphene modified with polyaniline	87.8 mF cm^{-2}	93%/10,000	EMITFSI/PVDF-HFP	0.22 mA cm^{-2}	[53]
MnO2@Au nanofiber	8.26 mF cm^{-2}	-/10,000	LiCl-PVA	5 mV s^{-1}	[54]
Carbon cloth-carbon fiber-TiO_2	270 mF cm^{-2}	-	1 M H_2SO_4	10 mA g^{-1}	[55]
2D-$LiCoO_2$	310 mF cm^{-2}	80.2%/2000	LiCl-PVA	5 mV s^{-1}	[56]
Carbon-doped titanium nitride	45.8 mF cm^{-2}	96%/5000	6 M KOH	10 mV s^{-1}	[57]
Nanoneedles-anchored CuCo-layered double hydroxide	7.02 mF cm^{-2}	96%/10,000	3 M KOH	30 mA cm^{-2}	[58]
Biochar	282 mF cm^{-2}	92%/1000	PVA solid-state film	5 mA cm^{-2}	This work
Biochar-PPy	370 mF cm^{-2}	72%/1000	PVA solid-state film	5 mA cm^{-2}	This work

The energy density and power density of the two SCs were obtained from Equations (2) and (3) and are shown in the Ragone plot (see Figure 7). As expected, the biochar-PPy-based electrode device showed the highest energy density. The PPy doping seemed to have increased its surface's polarity due to the presence of N-heteroatom functionalities, which boost electrostatic interactions between the electrode's active surface and the electrolyte ions, diminishing its equivalent resistance [59]. As expected, the improvement in the faradic redox reactions contributes to pseudocapacitance and enhances the supercapacitor's energy density due to the polypyrrole incorporation. The comparison with reported values in the literature is shown in the Ragone plot.

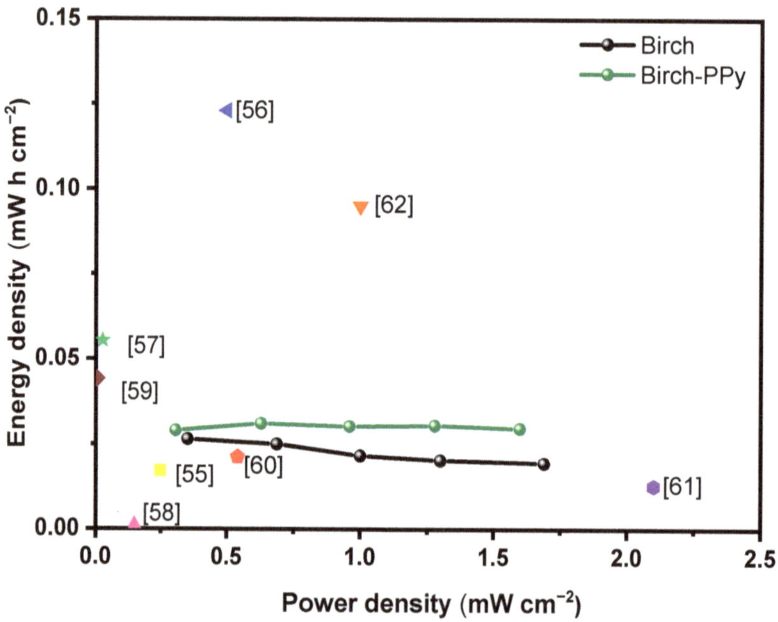

Figure 7. Ragone plot for biochar (curve in black) and biochar-PPy (curve in green) supercapacitors. For comparison, results from the literature are introduced in the colored dots: yellow [60], blue [61], green [62], magenta [63], wine [64], red [65], violet [66], and orange [67].

Another important characterization to be evaluated in supercapacitors is the measurement of the retention in the electrochemical properties of the materials under successive charge-discharge cycles. With this aim, successive charge-discharge cycles were imposed under two different currents (4 mA and 5 mA) in supercapacitors prepared with biochar (pristine) and doped by polypyrrole. After 1000 complete cycles of charge-discharge at a constant current density, Figure 8 summarizes the variation in the capacitance retention of the supercapacitors as a function of aging. In agreement with what was observed for carbonaceous materials, a slight linear decrease in the performance of the device is observed for a biochar-based supercapacitor with a value in the order of 93% retention in the electrochemical properties after 1000 cycles of use for both current densities. As expected, and as observed as a typical response for pseudocapacitive materials, reported for polypyrrole [68] and N-doping materials [17], a more pronounced decrease in the retention is observed for modified devices, which reach values in the order of 72% after 1000 cycles of use at 5 mA and 76.5% after 1000 cycles at 4 mA. Relative to the variation observed for these systems, it is worth observing that poor cycling performance is reported in the literature for polypyrrole-based supercapacitors [69], as a consequence of periodic contraction/expansion of polypyrrole chains under repeated anodic and cathodic polarization (under successive extraction/insertion of ions). These processes result in the formation of cracks that negatively affects the capacitance performance of the PPy-based devices. The incorporation of carbon-based materials minimizes the degradation rate since additives act as buffer space for polypyrrole volume variation.

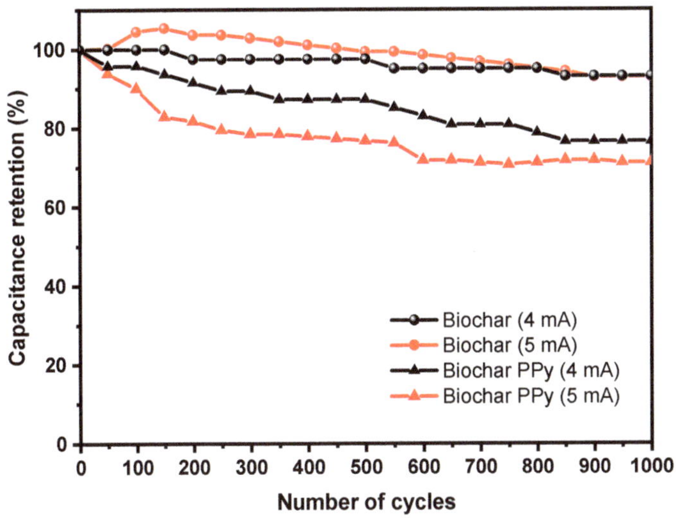

Figure 8. Comparison of retention in the capacitance after 1000 cycles of use of biochar at 4 mA (black dots), 5 mA (red dots), and biochar-PPy supercapacitors at 4 mA (black triangles) and 5 mA (red triangles).

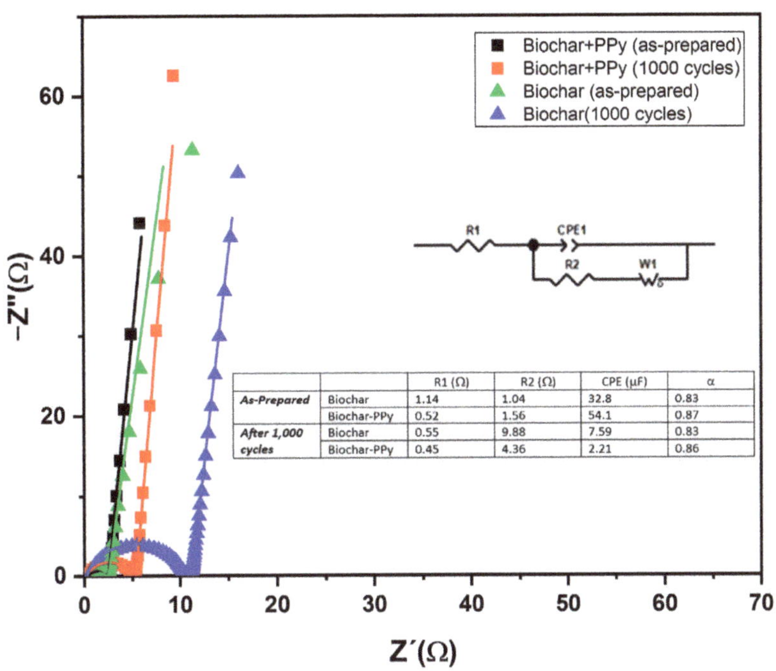

		R1 (Ω)	R2 (Ω)	CPE (μF)	α
As-Prepared	Biochar	1.14	1.04	32.8	0.83
	Biochar-PPy	0.52	1.56	54.1	0.87
After 1,000 cycles	Biochar	0.55	9.88	7.59	0.83
	Biochar-PPy	0.45	4.36	2.21	0.86

Figure 9. Nyquist diagram of both devices (0.01 Hz–1 MHz) for SCs composed with biochar and biochar-PPy samples characterized before and after the successive 1000 cycles of charge/discharge assays. The corresponding equivalent circuit (modified Randles circuit) and the corresponding fitting parameters for bulk resistance, charge transfer resistance, CPE, and parameter α are presented in the inset.

In addition, to further understand the intrinsic properties that influence the overall performance of the PPy-modified electrodes, electrical impedance spectroscopy (EIS) assays were conducted, in a frequency range from 0.01 Hz to 1 MHz, to further evaluate the behavior of biochar electrodes for supercapacitors (see Figure 9) before and after 1000 cycles of operation under conditions of high current (5 mA) and high degradation rate. The Nyquist plots of electrodes exhibit a semicircle and a straight line in the high and low-frequency regions, respectively, highlighting its capacitive behavior and ion diffusion [50]. The corresponding response for samples tested after 1000 cycles of charge/discharge shows a general shift to higher values of impedance in the diagram, confirming the degradation of the samples.

It is known that a longer diameter in the semicircle is attributed to a higher charge transfer resistance [40], and the biochar-PPy-based supercapacitor had a much smaller semi-circle, which reflected higher conductivity and low charge transfer resistance on the electrode structure, respectively; this leads to a faster ion transfer, which boosts the capacitance [70]. To quantitatively evaluate all of these aspects in the impedance spectrum, the overall spectrum was fitted by an equivalent circuit (shown in the inset of Figure 9) [71], known as the modified Randles circuit, that reproduces the straight line in the low frequency by the incorporation of a Warburg component in the circuit, and the depressed semicircle by a constant phase element (CPE), while resistors R_1 and R_2 reproduce the bulk resistance (R_1) and the charge transfer resistance (R_2–R_1).

The low values for R1 (see inset of Figure 9) confirm that both experimental systems (before and after continuous use) preserve low bulk resistance with prevailing resistance attributed to the biochar, which is minimally affected by the poor cycling performance of the conducting polymers. A general variation in the charge transfer resistance (R2) is observed for both experimental systems after successive uses, while an overall decrease in CPE value is observed with a stronger reduction in the response of the PPy-based device due to the characteristic lower cyclability attributed to pseudocapacitance-based materials

4. Conclusions

In this work, sustainable biomass electrode materials were prepared and employed in supercapacitors. A biomass-modified electrode with PPy was fully evaluated in terms of physicochemical and electrochemical features. The physicochemical characterization revealed that the PPy modification affected biochar morphology, specific surface area, pore structure, and the insertion of surface functionalities on the biochar's surface. The PPy loading reduced the electrode's specific surface area from 1052 $m^2\ g^{-1}$ to 87 $m^2\ g^{-1}$. Pseudocapacitive behavior prevails, with an improvement of 34% for areal capacitance. The wood-based electrodes delivered high areal capacitances of 282 and 370 mF cm^{-2} at 5 mA cm^{-2}, for pure biochar and biochar doped with PPy, respectively. At the end of 1000 cycles, the biochar doped with PPy had slightly reduced capacitance retention (72%) compared to pure biochar (92%). The improvement in the energy density and the areal capacitance for devices modified with polypyrrole confirm the relevant contribution of the conducting polymer on the overall pseudocapacitance behavior of the resulting supercapacitor.

Author Contributions: Conceptualization, R.M.A.P.L., G.S.d.R. and H.P.d.O.; Methodology, R.M.A.P.L. and G.S.d.R.; Validation, H.P.d.O.; Formal analysis, R.M.A.P.L. and G.S.d.R.; Investigation, R.M.A.P.L. and G.S.d.R.; Data curation, H.P.d.O.; Writing—original draft, R.M.A.P.L. and G.S.d.R.; Writing—review & editing, E.C.L., U.L., G.L.D.; Funding acquisition, E.C.L., U.L., G.L.D.; All authors have read and agreed to the published version of the manuscript.

Funding: Dr. dos Reis thanks Bio4Energy—a Strategic Research Environment appointed by the Swedish government and the Swedish University of Agricultural Sciences (SLU), for the funding support. This research was also funded by CAPES, FACEPE, FAPESB, and CNPq.

Data Availability Statement: Data can be made available upon request.

Acknowledgments: SLU, Bio4energy, CAPES, FACEPE, FAPESB, and CNPq. All authors thank the editors and anonymous reviewers for their valuable work.

Conflicts of Interest: The authors declare no conflict of interest.

References

1. Nirmaladevi, S.; Boopathiraja, R.; Kandasamy, S.K.; Sathishkumar, S.; Parthibavarman, M. Wood based biochar supported MnO_2 nanorods for high energy asymmetric supercapacitor applications. *Surfaces Interfaces* **2021**, *27*, 101548. [CrossRef]
2. Li, Y.; Zhu, L.; Shi, J.; Dou, Y.; Li, S.; You, R.; Zhang, S.; Miao, X.; Shi, S.; Ji, H.; et al. Super-hydrophilic microporous biochar from biowaste for supercapacitor application. *Appl. Surf. Sci.* **2021**, *561*, 150076. [CrossRef]
3. Khedulkar, A.P.; Dang, V.D.; Pandit, B.; Bui, T.A.N.; Tran, H.L.; Doong, R.-A. Flower-like nickel hydroxide@tea leaf-derived biochar composite for high-performance supercapacitor application. *J. Colloid Interface Sci.* **2022**, *623*, 845–855. [CrossRef] [PubMed]
4. Zhang, L.; Jiang, J.; Holm, N.; Chen, F. Mini-chunk biochar supercapacitors. *J. Appl. Electrochem.* **2014**, *44*, 1145–1151. [CrossRef]
5. Adhamash, E.; Pathak, R.; Qiao, Q.; Zhou, Y.; McTaggart, R. Gamma-radiated biochar carbon for improved supercapacitor performance. *RSC Adv.* **2020**, *10*, 29910–29917. [CrossRef]
6. Husain, Z.; Raheman, A.S.; Ansari, K.B.; Pandit, A.B.; Khan, M.S.; Qyyum, M.A.; Lam, S.S. Nano-sized mesoporous biochar derived from biomass pyrolysis as electrochemical energy storage supercapacitor. *Mater. Sci. Energy Technol.* **2022**, *5*, 99–109. [CrossRef]
7. Gupta, R.K.; Dubey, M.; Kharel, P.; Gu, Z.; Fan, Q.H. Biochar activated by oxygen plasma for supercapacitors. *J. Power Sources* **2015**, *274*, 1300–1305. [CrossRef]
8. Thines, K.; Abdullah, E.; Mubarak, N.; Ruthiraan, M. In-situ polymerization of magnetic biochar—Polypyrrole composite: A novel application in supercapacitor. *Biomass Bioenergy* **2017**, *98*, 95–111. [CrossRef]
9. Jiang, J.; Zhang, L.; Wang, X.; Holm, N.; Rajagopalan, K.; Chen, F.; Ma, S. Highly ordered macroporous woody biochar with ultra-high carbon content as supercapacitor electrodes. *Electrochim. Acta* **2013**, *113*, 481–489. [CrossRef]
10. Oliveira, G.d.S.; Candido, I.C.M.; Lima, R.M.A.P.; de Oliveira, H.P. All-In-One Energy Harvesting/Storage Integrated Systems Based on Eggshell Membranes. *ACS Appl. Electron. Mater.* **2022**, *4*, 4708–4718. [CrossRef]
11. Wu, Z.; Li, L.; Yan, J.-M.; Zhang, X.-B. Materials Design and System Construction for Conventional and New-Concept Supercapacitors. *Adv. Sci.* **2017**, *4*, 1600382. [CrossRef] [PubMed]
12. Vatansever, D.; Haciismailoglu, M. Pencil graphite/polypyrrole supercapacitors: The structural and chemical changes of the electrodes after capacitive measurements. *Mater. Chem. Phys.* **2023**, *295*, 127073. [CrossRef]
13. Jiang, H.; Sun, T.; Li, C.; Ma, J. Hierarchical porous nanostructures assembled from ultrathin MnO_2 nanoflakes with enhanced supercapacitive performances. *J. Mater. Chem.* **2012**, *22*, 2751–2756. [CrossRef]
14. Ji, W.; Ji, J.; Cui, X.; Chen, J.; Liu, D.; Deng, H.; Fu, Q. Polypyrrole encapsulation on flower-like porous NiO for advanced high-performance supercapacitors. *Chem. Commun.* **2015**, *51*, 7669–7672. [CrossRef] [PubMed]
15. Liu, Y.; Peng, X. Recent advances of supercapacitors based on two-dimensional materials. *Appl. Mater. Today* **2017**, *8*, 104–115. [CrossRef]
16. Baboukani, A.R.; Khakpour, I.; Drozd, V.; Allagui, A.; Wang, C. Single-step exfoliation of black phosphorus and deposition of phosphorene *via* bipolar electrochemistry for capacitive energy storage application. *J. Mater. Chem. A* **2019**, *7*, 25548–25556. [CrossRef]
17. Orlando, J.D.; Lima, R.M.A.P.; Li, L.; Sydlik, S.A.; de Oliveira, H.P. Electrochemical Performance of N-Doped Carbon-Based Electrodes for Supercapacitors. *ACS Appl. Electron. Mater.* **2022**, *4*, 5040–5054. [CrossRef]
18. Jiang, Y.; Liu, Z.; Lv, Y.; Tang, A.; Dai, L.; Wang, L.; He, Z. Perovskite enables high performance vanadium redox flow battery. *Chem. Eng. J.* **2022**, *443*, 136341. [CrossRef]
19. Mao, C.; Chang, Y.; Zhao, X.; Dong, X.; Geng, Y.; Zhang, N.; Dai, L.; Wu, X.; Wang, L.; He, Z. Functional carbon materials for high-performance Zn metal anodes. *J. Energy Chem.* **2022**, *75*, 135–153. [CrossRef]
20. Dos Reis, G.S.; Larsson, S.H.; Mathieu, M.; Thyrel, M.; Pham, T.N. Application of design of experiments (DoE) for optimised production of micro- and mesoporous Norway spruce bark activated carbons. *Biomass Convers. Biorefinery* **2021**, *1*, 1–19. [CrossRef]
21. Saini, S.; Chand, P.; Joshi, A. Biomass derived carbon for supercapacitor applications: Review. *J. Energy Storage* **2021**, *39*, 102646. [CrossRef]
22. González-Hourcade, M.; dos Reis, G.S.; Grimm, A.; Dinh, V.M.; Lima, E.C.; Larsson, S.H.; Gentili, F.G. Microalgae biomass as a sustainable precursor to produce nitrogen-doped biochar for efficient removal of emerging pollutants from aqueous media. *J. Clean. Prod.* **2022**, *348*, 131280. [CrossRef]
23. dos Reis, G.S.; Guy, M.; Mathieu, M.; Jebrane, M.; Lima, E.C.; Thyrel, M.; Dotto, G.L.; Larsson, S.H. A comparative study of chemical treatment by MgCl2, ZnSO4, ZnCl2, and KOH on physicochemical properties and acetaminophen adsorption performance of biobased porous materials from tree bark residues. *Colloids Surfaces A Physicochem. Eng. Asp.* **2022**, *642*, 128626. [CrossRef]

24. Guy, M.; Mathieu, M.; Anastopoulos, I.P.; Martínez, M.G.; Rousseau, F.; Dotto, G.L.; de Oliveira, H.P.; Lima, E.C.; Thyrel, M.; Larsson, S.H.; et al. Process Parameters Optimization, Characterization, and Application of KOH-Activated Norway Spruce Bark Graphitic Biochars for Efficient Azo Dye Adsorption. *Molecules* **2022**, *27*, 456. [CrossRef] [PubMed]
25. Le, T.-H.; Kim, Y.; Yoon, H. Electrical and Electrochemical Properties of Conducting Polymers. *Polymers* **2017**, *9*, 150. [CrossRef] [PubMed]
26. Kim, S.; Jang, L.K.; Park, H.S.; Lee, J.Y. Electrochemical deposition of conductive and adhesive polypyrrole-dopamine films. *Sci. Rep.* **2016**, *6*, 30475. [CrossRef]
27. Cao, Y.; Qi, X.; Hu, K.; Wang, Y.; Gan, Z.; Li, Y.; Hu, G.; Peng, Z.; Du, K. Conductive Polymers Encapsulation To Enhance Electrochemical Performance of Ni-Rich Cathode Materials for Li-Ion Batteries. *ACS Appl. Mater. Interfaces* **2018**, *10*, 18270–18280. [CrossRef]
28. de Oliveira, H.P.; Sydlik, S.A.; Swager, T.M. Supercapacitors from Free-Standing Polypyrrole/Graphene Nanocomposites. *J. Phys. Chem. C* **2013**, *117*, 10270–10276. [CrossRef]
29. Alcaraz-Espinoza, J.J.; De Melo, C.P.; De Oliveira, H.P. Fabrication of Highly Flexible Hierarchical Polypyrrole/Carbon Nanotube on Eggshell Membranes for Supercapacitors. *ACS Omega* **2017**, *2*, 2866–2877. [CrossRef]
30. Candido, I.C.M.; Oliveira, G.d.S.; Viana, G.G.; da Silva, F.A.G.; da Costa, M.M.; de Oliveira, H.P. Wearable Triboelectric Nanogenerators Based on Chemical Modification of Conventional Textiles for Application in Electrically Driven Antibacterial Devices. *ACS Appl. Electron. Mater.* **2022**, *4*, 334–344. [CrossRef]
31. Barakzehi, M.; Montazer, M.; Sharif, F.; Norby, T.; Chatzitakis, A. A textile-based wearable supercapacitor using reduced graphene oxide/polypyrrole composite. *Electrochim. Acta* **2019**, *305*, 187–196. [CrossRef]
32. Soleimani, M.; Ghorbani, M.; Salahi, S. Antibacterial Activity of Polypyrrole-Chitosan Nanocomposite: Mechanism of Action. *Int. J. Nanosci. Nanotechnol.* **2016**, *12*, 191–197.
33. Gao, M.; Wang, W.-K.; Zheng, Y.-M.; Zhao, Q.-B.; Yu, H.-Q. Hierarchically porous biochar for supercapacitor and electrochemical H_2O_2 production. *Chem. Eng. J.* **2020**, *402*, 126171. [CrossRef]
34. Ma, Z.-W.; Liu, H.-Q.; Lü, Q.-F. Porous biochar derived from tea saponin for supercapacitor electrode: Effect of preparation technique. *J. Energy Storage* **2021**, *40*, 102773. [CrossRef]
35. da Silva, R.J.; Lima, R.M.A.P.; de Oliveira, M.C.A.; Alcaraz-Espinoza, J.J.; de Melo, C.P.; de Oliveira, H.P. Supercapacitors Based on (Carbon Nanostructure)/PEDOT/(Eggshell Membrane) Electrodes. *J. Electroanal. Chem.* **2020**, *856*, 113658. [CrossRef]
36. Lima, R.M.A.P.; Alcaraz-Espinoza, J.J.; da Silva, F.A.G.; de Oliveira, H.P. Multifunctional Wearable Electronic Textiles Using Cotton Fibers with Polypyrrole and Carbon Nanotubes. *ACS Appl. Mater. Interfaces* **2018**, *10*, 13783–13795. [CrossRef] [PubMed]
37. Alcaraz-Espinoza, J.J.; de Oliveira, H.P. Flexible Supercapacitors Based on a Ternary Composite of Polyaniline/Polypyrrole/Graphite on Gold Coated Sandpaper. *Electrochim. Acta* **2018**, *274*, 200–207. [CrossRef]
38. Lima, D.R.; Lima, E.C.; Thue, P.S.; Dias, S.L.; Machado, F.M.; Seliem, M.K.; Sher, F.; dos Reis, G.S.; Saeb, M.R.; Rinklebe, J. Comparison of acidic leaching using a conventional and ultrasound-assisted method for preparation of magnetic-activated biochar. *J. Environ. Chem. Eng.* **2021**, *9*, 105865. [CrossRef]
39. Lima, R.M.A.P.; dos Reis, G.S.; Thyrel, M.; Alcaraz-Espinoza, J.J.; Larsson, S.H.; de Oliveira, H.P. Facile Synthesis of Sustainable Biomass-Derived Porous Biochars as Promising Electrode Materials for High-Performance Supercapacitor Applications. *Nanomaterials* **2022**, *12*, 866. [CrossRef]
40. dos Reis, G.S.; Lima, R.M.A.P.; Larsson, S.H.; Subramaniyam, C.M.; Dinh, V.M.; Thyrel, M.; de Oliveira, H.P. Flexible supercapacitors of biomass-based activated carbon-polypyrrole on eggshell membranes. *J. Environ. Chem. Eng.* **2021**, *9*, 106155. [CrossRef]
41. Thommes, M.; Kaneko, K.; Neimark, A.V.; Olivier, J.P.; Rodriguez-Reinoso, F.; Rouquerol, J.; Sing, K.S.W. Physisorption of gases, with special reference to the evaluation of surface area and pore size distribution (IUPAC Technical Report). *Pure Appl. Chem.* **2015**, *87*, 1051–1069. [CrossRef]
42. dos Reis, G.S.; Bergna, D.; Tuomikoski, S.; Grimm, A.; Lima, E.C.; Thyrel, M.; Skoglund, N.; Lassi, U.; Larsson, S.H. Preparation and Characterization of Pulp and Paper Mill Sludge-Activated Biochars Using Alkaline Activation: A Box–Behnken Design Approach. *ACS Omega* **2022**, *7*, 32620–32630. [CrossRef] [PubMed]
43. Zhou, H.; Han, G.; Xiao, Y.; Chang, Y.; Zhai, H.-J. Facile preparation of polypyrrole/graphene oxide nanocomposites with large areal capacitance using electrochemical codeposition for supercapacitors. *J. Power Sources* **2014**, *263*, 259–267. [CrossRef]
44. Jyothibasu, J.P.; Chen, M.-Z.; Lee, R.-H. Polypyrrole/Carbon Nanotube Freestanding Electrode with Excellent Electrochemical Properties for High-Performance All-Solid-State Supercapacitors. *ACS Omega* **2020**, *5*, 6441–6451. [CrossRef] [PubMed]
45. Dubal, D.P.; Lee, S.H.; Kim, J.G.; Kim, W.B.; Lokhande, C.D. Porous polypyrrole clusters prepared by electropolymerization for a high performance supercapacitor. *J. Mater. Chem.* **2012**, *22*, 3044–3052. [CrossRef]
46. Huang, T.; Zheng, B.; Kou, L.; Gopalsamy, K.; Xu, Z.; Gao, C.; Meng, Y.; Wei, Z. Flexible high performance wet-spun graphene fiber supercapacitors. *RSC Adv.* **2013**, *3*, 23957–23962. [CrossRef]
47. Liu, Y.; Wang, N.; Yang, C.; Hu, W. Sol–gel synthesis of nanoporous $NiCo_2O_4$ thin films on ITO glass as high-performance supercapacitor electrodes. *Ceram. Int.* **2016**, *42*, 11411–11416. [CrossRef]
48. Zhang, H.; Wei, J.; Yan, Y.; Guo, Q.; Xie, L.; Yang, Z.; He, J.; Qi, W.; Cao, Z.; Zhao, X.; et al. Facile and Scalable Fabrication of MnO_2 Nanocrystallines and Enhanced Electrochemical Performance of MnO_2/MoS_2 Inner Heterojunction Structure for Supercapacitor Application. *J. Power Sources* **2020**, *450*, 227616. [CrossRef]

49. Zhang, J.; Wang, Y.; Wu, J.; Shu, X.; Yu, C.; Cui, J.; Qin, Y.; Zhang, Y.; Ajayan, P.M.; Wu, Y. Remarkable supercapacitive performance of TiO_2 nanotube arrays by introduction of oxygen vacancies. *Chem. Eng. J.* **2017**, *313*, 1071–1081. [CrossRef]
50. Prasad, K.; Reddy, G.R.; Rajesh, M.; Babu, P.R.; Shanmugam, G.; Sushma, N.J.; Reddy, M.S.P.; Raju, B.D.P.; Mallikarjuna, K. Electrochemical Performance of 2D-Hierarchical Sheet-Like $ZnCo_2O_4$ Microstructures for Supercapacitor Applications. *Crystals* **2020**, *10*, 566. [CrossRef]
51. Li, Q.; Cheng, H.; Wu, X.; Wang, C.-F.; Wu, G.; Chen, S. Enriched carbon dots/graphene microfibers towards high-performance micro-supercapacitors. *J. Mater. Chem. A* **2018**, *6*, 14112–14119. [CrossRef]
52. Chen, Q.; Meng, Y.; Hu, C.; Zhao, Y.; Shao, H.; Chen, N.; Qu, L. MnO_2-modified hierarchical graphene fiber electrochemical supercapacitor. *J. Power Sources* **2014**, *247*, 32–39. [CrossRef]
53. Zhang, M.; Wang, X.; Yang, T.; Zhang, P.; Wei, X.; Zhang, L.; Li, H. Polyaniline/graphene hybrid fibers as electrodes for flexible supercapacitors. *Synth. Met.* **2020**, *268*, 116484. [CrossRef]
54. Park, J.; An, G.-H. Interface-engineered electrode and electrolyte for the improved energy-storing performance and stable mechanical flexibility of fibrous supercapacitors. *Appl. Surf. Sci.* **2021**, *549*, 149326. [CrossRef]
55. Lal, M.S.; Ramaprabhu, S. High Areal Capacitance of Flexible Supercapacitors Fabricated with Carbon Cloth-Carbon Fiber-TiO_2 Electrodes and Different Hydrogel Polymer Electrolytes. *J. Electrochem. Soc.* **2022**, *169*, 20514. [CrossRef]
56. Lu, J.; Ran, H.; Li, J.; Wan, J.; Wang, C.; Ji, P.; Wang, X.; Liu, G.; Hu, C. A Fast Composite-Hydroxide-Mediated Approach for Synthesis of 2D-$LiCoO_2$ for High Performance Asymmetric Supercapacitor. *Electrochim. Acta* **2020**, *331*, 135426. [CrossRef]
57. Shi, J.; Jiang, B.; Li, C.; Liu, Z.; Yan, F.; Liu, X.; Li, H.; Yang, C.; Dong, D.; Hao, J. Study on capacitance properties of the sputtered carbon doped titanium nitride electrode material for supercapacitor. *Vacuum* **2022**, *198*, 110893. [CrossRef]
58. Mane, S.M.; Teli, A.M.; Yang, H.K.; Kwon, E.; Nimbalkar, N.A.; Patil, D.R.; Shin, J.C. Nanoneedles anchored ultrathin petals of CuCo layered double hydroxide with high areal capacitance and long cycle life for high-performance hybrid supercapacitors. *J. Energy Storage* **2023**, *62*, 106941. [CrossRef]
59. Rustamaji, H.; Prakoso, T.; Devianto, H.; Widiatmoko, P.; Kurnia, K.A. Facile synthesis of N, S-modified activated carbon from biomass residue for promising supercapacitor electrode applications. *Bioresour. Technol. Rep.* **2023**, *21*, 101301. [CrossRef]
60. Li, K.; Liu, X.; Chen, S.; Pan, W.; Zhang, J. A flexible solid-state supercapacitor based on graphene/polyaniline paper electrodes. *J. Energy Chem.* **2019**, *32*, 166–173. [CrossRef]
61. Chen, C.; Qin, H.; Cong, H.; Yu, S. A Highly Stretchable and Real-Time Healable Supercapacitor. *Adv. Mater.* **2019**, *31*, e1900573. [CrossRef] [PubMed]
62. Wang, W.; Yang, Y.; Chen, Z.; Deng, Z.; Fan, L.; Guo, W.; Xu, J.; Meng, Z. High-performance yarn supercapacitor based on directly twisted carbon nanotube@bacterial cellulose membrane. *Cellulose* **2020**, *27*, 7649–7661. [CrossRef]
63. Sim, H.J.; Choi, C.; Lee, D.Y.; Kim, H.; Yun, J.-H.; Kim, J.M.; Kang, T.M.; Ovalle, R.; Baughman, R.H.; Kee, C.W.; et al. Biomolecule based fiber supercapacitor for implantable device. *Nano Energy* **2018**, *47*, 385–392. [CrossRef]
64. Yang, Z.; Zhao, W.; Niu, Y.; Zhang, Y.; Wang, L.; Zhang, W.; Xiang, X.; Li, Q. Direct spinning of high-performance graphene fiber supercapacitor with a three-ply core-sheath structure. *Carbon* **2018**, *132*, 241–248. [CrossRef]
65. Zhang, W.; Guo, R.; Sun, J.; Dang, L.; Liu, Z.; Lei, Z.; Sun, Q. Textile carbon network with enhanced areal capacitance prepared by chemical activation of cotton cloth. *J. Colloid Interface Sci.* **2019**, *553*, 705–712. [CrossRef]
66. Qi, K.; Hou, R.; Zaman, S.; Qiu, Y.; Xia, B.Y.; Duan, H. Construction of Metal–Organic Framework/Conductive Polymer Hybrid for All-Solid-State Fabric Supercapacitor. *ACS Appl. Mater. Interfaces* **2018**, *10*, 18021–18028. [CrossRef]
67. Bhargava, P.; Liu, W.; Pope, M.; Tsui, T.; Yu, A. Substrate comparison for polypyrrole-graphene based high-performance flexible supercapacitors. *Electrochim. Acta* **2020**, *358*, 136846. [CrossRef]
68. Moreno Araújo Pinheiro Lima, R.; de Oliveira, H.P. Carbon Dots Reinforced Polypyrrole/ Graphene Nanoplatelets on Flexible Eggshell Membranes as Electrodes of All-Solid Flexible Supercapacitors. *J. Energy Storage* **2020**, *28*, 101284. [CrossRef]
69. Hryniewicz, B.M.; Lima, R.V.; Marchesi, L.F.; Vidotti, M. Impedimetric studies about the degradation of polypyrrole nanotubes during galvanostatic charge and discharge cycles. *J. Electroanal. Chem.* **2019**, *855*, 113636. [CrossRef]
70. Soram, B.S.; Dai, J.; Kshetri, T.; Kim, N.H.; Lee, J.H. Vertically grown and intertwined $Co(OH)_2$ nanosheet@Ni-mesh network for transparent flexible supercapacitor. *Chem. Eng. J.* **2020**, *391*, 123540. [CrossRef]
71. Baboukani, A.R.; Khakpour, I.; Adelowo, E.; Drozd, V.; Shang, W.; Wang, C. High-performance red phosphorus-sulfurized polyacrylonitrile composite by electrostatic spray deposition for lithium-ion batteries. *Electrochim. Acta* **2020**, *345*, 136227. [CrossRef]

C Journal of Carbon Research

MDPI

Article

Density Functional Theory Analysis of the Impact of Boron Concentration and Surface Oxidation in Boron-Doped Graphene for Sodium and Aluminum Storage

Milica S. Ritopečki [1], Natalia V. Skorodumova [2,3], Ana S. Dobrota [1,*] and Igor A. Pašti [1,*]

[1] Faculty of Physical Chemistry, University of Belgrade, Studentski Trg 12-16, 11158 Belgrade, Serbia; milicaritopecki@gmail.com
[2] Department of Materials Science and Engineering, School of Industrial Engineering and Management, KTH—Royal Institute of Technology, Brinellvägen 23, 10044 Stockholm, Sweden; snv123@kth.se
[3] Applied Physics, Division of Materials Science, Department of Engineering Sciences and Mathematics, Luleå University of Technology, 97187 Luleå, Sweden
* Correspondence: ana.dobrota@ffh.bg.ac.rs (A.S.D.); igor@ffh.bg.ac.rs (I.A.P.)

Abstract: Graphene is thought to be a promising material for many applications. However, pristine graphene is not suitable for most electrochemical devices, where defect engineering is crucial for its performance. We demonstrate how the boron doping of graphene can alter its reactivity, electrical conductivity and potential application for sodium and aluminum storage, with an emphasis on novel metal-ion batteries. Using Density Functional Theory calculations, we investigate both the influence of boron concentration and the oxidation of the material on the mentioned properties. It is demonstrated that the presence of boron in graphene increases its reactivity towards atomic hydrogen and oxygen-containing species; in other words, it makes B-doped graphene more prone to oxidation. Additionally, the presence of these surface functional groups significantly alters the type and strength of the interaction of Na and Al with the given materials. Boron-doping and the oxidation of graphene is found to increase the Na storage capacity of graphene by a factor of up to four, and the calculated sodiation potentials indicate the possibility of using these materials as electrode materials in high-voltage Na-ion batteries.

Keywords: graphene; boron-doped graphene; reactivity; oxidation; metal-ion batteries

1. Introduction

Graphene is a two-dimensional material with a graphite monolayer structure and sp^2-hybridized carbon (C) atoms arranged in a honeycomb lattice [1,2]. Due to its rare properties, such as good electric conductivity and a large surface area, graphene has attracted interest as a potential anode material. However, the fact that it is a zero-bandgap semiconductor, along with its chemical inertness, limits its application possibilities [3–7]. Therefore, the graphene lattice is usually changed by chemical doping or by introducing covalent bonds with certain chemical groups or even molecules [8,9]. It should, however, be noted that the defects in graphene can also occur spontaneously during synthesis, and they can be sometimes difficult to predict and control [10]. The simplest modification method of pristine is chemical doping by heteroatoms, which improves the properties of pristine graphene by improving the metal–surface interaction, adsorption and charge transfer abilities [2,4,11,12]. Heteroatoms such as boron (B), nitrogen (N), sulfur (S), fluorine (F), and phosphorus (P) are frequently chosen for this purpose since they change the bandgap and the position of the Fermi level, which depends on their concentration [4,13].

Carbon-based materials, including graphene, are widely investigated as candidates for electrode materials in metal-ion batteries (MIBs). For MIBs, the strength and the nature of the interactions of the metal (and/or metal-based ions) with electrode materials are essential. Novel types of MIBs focus on metals other than Li as the electroactive species due to the problems associated with Li-ion batteries [14] and Li abundance in general. These other metals include Na and Al, as they have small a mass and radius and a high abundance in Earth's crust [15]. Additionally, further advantages of Al are its three-electron redox property and low flammability. However, both Na and Al interact with pristine graphene relatively weakly [16], resulting in low voltages in hypothetical novel MIBs with graphene as the electrode material. Thus, chemical functionalization is needed to boost graphene's Na and Al storage capacity. Some reports of Al- and Na-ion batteries, which use carbon-based electrodes, can already be found in the literature [17–20].

Boron-doped graphene is an especially good candidate for the doping of graphene since B- and C-atoms have similar radii, making a C-B bond in a monolayer only ~0.5% longer than the C-C one. Because of this, doping does not affect the planarity of the graphene sheet or the sp^2 hybridization of the carbon atoms within [11,21]. Boron is less electronegative than carbon, which causes a difference in electron density between boron and carbon atoms, leaving the boron site with an electron deficit [22]. In addition, since boron has one electron less, B-graphene acts as a p-type semiconductor [23]. B-graphene also has, similarly to pristine graphene, a very low or nondetectable total magnetization [12,23]. For now, B-graphene has primarily found its place in energy-related applications, such as batteries and supercapacitors [24,25], but there are also studies where it is used as a gas and optical sensor material [26,27] and in the biomedical field of study [28]. For example, Ling and Mizuno used first-principle calculations to demonstrate that the sodiation of boron-doped graphene preserves its structural integrity, avoids the formation of dendrites and allows for approx. 2 times the capacity of the graphite anode in Li-IB and approx. 2.5 times higher than hard carbon in Na-IB [29]. Zhang et al. presented the facile synthesis of boron-doped reduced graphene oxide, which can be used as a sodium-ion battery anode [30]. Recently, Yiqun et al. demonstrated that the cathode based on boron-doped reduced graphene oxide displays a high Al-storage capacity and outstanding long-term stability [31].

In this study, we continue our previous work [32,33], where it was shown that the introduction of dopants (B, N, P or S) into oxidized graphene or the controlled oxidation of doped graphene can enhance the Na storage capabilities of the material, with boron found to be a very suitable dopant for this purpose. Thus, we investigate the effects of different concentrations of boron as a dopant in graphene on its properties, which are not addressed in detail in the literature but are of high importance for the electrochemical applications of B-doped graphene. First, we focus on the reactivity of the material towards H, O and OH, and next, we investigate the adsorption of Na and Al onto bare and oxidized boron-doped graphene to explore the possibility of using such modified graphene surfaces in novel energy storage devices.

2. Methods

In order to carry out Density Functional Theory (DFT) calculations, the open-source program packet Quantum ESSPRESO [34,35] was used. GGA and PBE functionals were used as part of this package. Since the adsorption of sodium and aluminum on graphene-based materials was investigated, dispersion interactions need to be included by utilizing the DFT-D2 correction [36]. The kinetic energy cut-off of plane waves was set at 36 Ry, and the density cut-off was set at 576 Ry. Spin polarization was included in all the conducted calculations. The first irreducible Brillouin zone was obtained using a Γ-centered $4 \times 4 \times 1$ grid utilizing the general Monkhorst–Pack scheme [37]. Löwdin charges were used for discussing charge transfer. To investigate electronic structure, a denser centered $20 \times 20 \times 1$ grid was used.

Pristine graphene was modelled as a $(3\sqrt{3} \times 3\sqrt{3})R30^\circ$ structure containing 54 carbon atoms, C_{54}. Boron-doped graphene was constructed by replacing one, two or three carbon atoms with boron, obtaining graphene with 1.85, 3.70 or 5.56 at.% of B, respectively. To evaluate the stability of boron-doped graphene, the binding energy (E_b) of boron at the vacancy sites was used. The binding energy of the nth ($n \in \{1, 2, 3\}$) substitutionally introduced boron atom is calculated as

$$E_b(n\text{th B}) = E(C_{54-n}B_n) - E(C_{54-n}B_{n-1}) - E(B) \quad (1)$$

where $E(C_{54-n}B_n)$ stands for the total energy of graphene doped with n atoms of B, $E(C_{54-n}B_{n-1})$ is the total energy of graphene doped with $n - 1$ atoms of B and a vacancy at the site where the nth B will be introduced and E(B) the total energy of an isolated boron atom. Since the adsorption of atomic hydrogen, atomic oxygen and hydroxyl groups on $C_{54-n}B_n$ was also investigated, their corresponding adsorption energies (E_{ads}) were calculated using the following formula:

$$E_{ads}(A) = E_{subs+A} - E_{subs} - E_A \quad (2)$$

where E_{subs+A}, E_{subs} and E_A stand for the total energies of the optimized adsorbate–substrate system, bare substrate ($C_{54-n}B_n$), and isolated adsorbate atom/group (A = H, O, or OH), respectively. The most stable O@$C_{54-n}B_n$ and OH@$C_{54-n}B_n$ structures were further used as models of oxidized $C_{54-n}B_n$.

Finally, the adsorption of Na and Al on bare $C_{54-n}B_n$ and its oxidized forms (O@$C_{54-n}B_n$ and OH@$C_{54-n}B_n$) was investigated, while the strength of their interaction with the substrate was estimated in terms of differential and integral adsorption energies. The differential adsorption energy ($E_{ads,diff}$) was calculated as

$$E_{ads,diff}\,(M) = E_{subs+mM} - E_{subs+(m-1)M} - E_M \quad (3)$$

where "subs" can be $C_{54-n}B_n$, O@$C_{54-n}B_n$ or OH@$C_{54-n}B_n$, while $E_{subs+mM}$ stands for the total energy of the chosen substrate with m atoms of metal M adsorbed (i.e., m is the number of atoms M, M = Na or Al). Similarly, $E_{subs+(m-1)M}$ stands for the total energy of the chosen substrate with $m - 1$ atoms of metal M adsorbed. The integral adsorption energy ($E_{ads,int}$) was calculated as

$$E_{ads,int}\,(M) = (E_{subs+mM} - E_{subs} - m \cdot E_M)/m \quad (4)$$

where E_{subs} stands for the total energy of the optimized substrate without any M adsorbed. Charge redistribution caused by metal interaction with the chosen model systems was investigated through a 3D plot of charge difference ($\Delta\rho$), defined as:

$$\Delta\rho = \rho_{subs+M} - \rho_{subs,frozen} - \rho_M \quad (5)$$

where ρ_{subs+M}, $\rho_{subs,frozen}$ and ρ_M stand for the ground state charge densities of the substrate interacting with M, the ground state charge densities of the substrate when M is removed (with frozen geometry) and that of the isolated M atom, respectively.

Graphical representations of all the graphene structures in this paper were made using VESTA [38].

3. Results

3.1. Graphene Doping by Boron

Pristine graphene, C_{54}, was used as a starting model. Its optimization resulted in a C-C bond length of 1.43 Å, which is in agreement with previous studies [12,39,40]. B-doped graphene was modelled by replacing n carbon atoms by boron, as described in the Computational Details section. While the position of the first boron atom can be chosen arbitrarily, it is necessary to perform additional calculations to find the optimal positions of the second and the third boron atom relative to the first one, or the first and second one, respectively. In the case of $C_{52}B_2$ systems, we find that the energetically least stable configuration is the one where the two B atoms are next to each other (Figure 1). The systems become more stable as the two B atoms separate, and the optimal structure corresponds to the state where they are one across from the other, on opposite sides of the same C_4B_2 hexagon (Figure 1). We note that substitutional B atom clustering was recently observed for B-doped graphene single layers on the Ir(111) surface [41]. Boron atoms did not prefer direct bonding, that is, the configuration in which they are adjacent to each other in a C_4H_2 ring. The situation we find is not identical to that described in ref. [41], which also included the geometric deformation of the B-doped graphene sheet caused by its bending towards the Ir(111) surface.

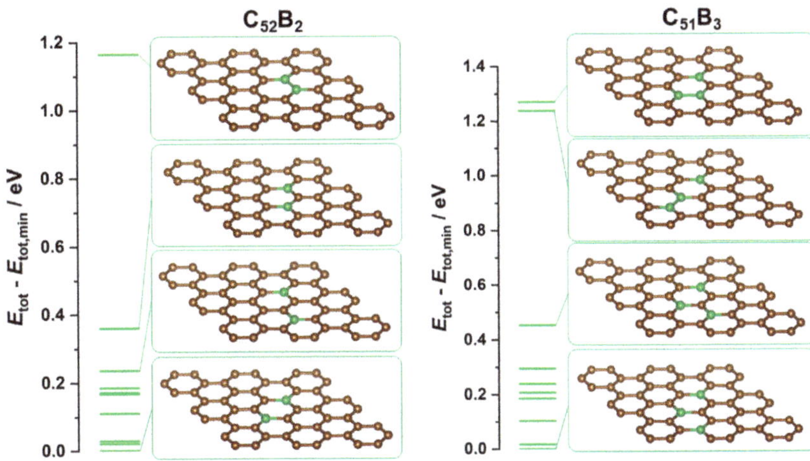

Figure 1. Relative stabilities (in terms of total energies, E_{tot}) of the investigated $C_{52}B_2$ (**left**) and $C_{51}B_3$ (**right**) structures compared to the most stable one with $E_{tot,min}$, which corresponds to $E_{tot} - E_{tot,min} = 0$.

A similar pattern was found upon the introduction of the third boron atom—the dopant atoms are one across from the other on opposite sides of one hexagon, while it is optimal to have two boron atoms inside one hexagon. The first boron atom (the case of $C_{53}B$) is found to be incorporated into the graphene lattice with a binding energy of −12.90 eV and a C-B bond length of 1.49 Å, which is in agreement with our previous studies [12,42]. When more B atoms are introduced, the binding energies follow the following trend: E_b(1st B) < E_b(2nd B) < E_b(3rd B), i.e., the first B is bound the strongest. The differences in their E_b are up to 6.2%. The results are summarized in Table 1. The results presented here suggest that boron atoms at higher doping levels could form a 2D array even in a self-standing B-doped graphene sheet, similarly to B-doped graphene on Ir(111) [41]. All the investigated $C_{54-n}B_n$ systems were found to be nonmagnetic. We observed that there was no disruption to the planarity, even with the addition of three B atoms, which is a consequence of the similar atomic radii of B and C. Upon the introduction of boron, the Bader charge of its first neighbouring carbon atoms increased by 0.1 e. On the other hand, the charge of boron atoms decreased by 0.1–0.2 e, which is expected since boron is less electronegative than carbon.

Table 1. Binding energies of boron atom (E_b(B)), C-B bond lengths (d(C-B)) and partial Löwdin charges of B (Δq(B)) for the studied $C_{54-n}B_n$ models.

Model	E_b(B)/eV	d(C-B)/Å	Δq(B)/e
$C_{53}B$	−12.90	1.49	−0.1
$C_{52}B_2$	−12.46	1.51	−0.2
$C_{51}B_3$	−12.15	1.51	−0.2

Upon boron insertion into a graphene lattice, the Fermi energy is shifted towards higher energies. In Figure S3, the electronic structures of pristine graphene and doped species were also compared. The energy range in this figure is chosen so that the p states of boron atoms can be clearly seen. For the full range DOS plots please see Figure S3, right. It is well known that pristine graphene is a semimetal with a zero-band gap [23]. This was also observed here, and it is also shown that doped models behave as conductors. This is very important for the materials' potential electrochemical purposes.

After setting the models of boron-doped graphene with various concentrations of B, we move on to exploring their reactivity towards species of interest.

3.2. Reactivity and Oxidation

The H, O and OH adsorption trends of $C_{54-n}B_n$ are summarized in Figure 2. When investigating H adsorption (Figure 2, top), the C-top sites were found to be the most favourable ones. For the B-doped surfaces, the carbon atom that is the closest to a boron atom is the one where H prefers to adsorb. This is in agreement with some previous research [12,43]. We find that with the increment of boron percentage in the system, the H adsorption energy decreases, i.e., H is bound more strongly. For pristine graphene, the energy of H adsorption is −0.79 eV, while for $C_{51}B_3$, it is −2.06 eV. The C-H bond length does not change significantly upon B introduction (1.13–1.15 Å).

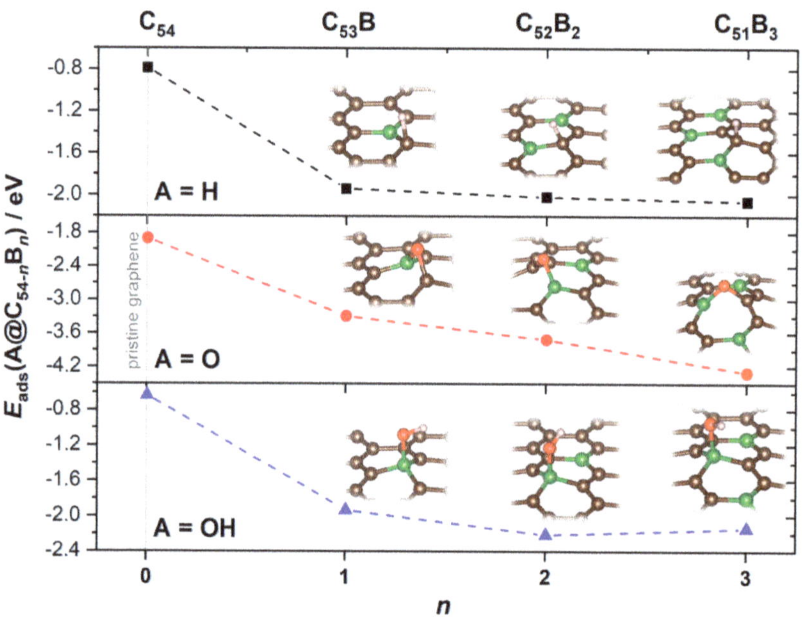

Figure 2. The adsorption trends of H, O and OH on $C_{54-n}B_n$ models: the relationship between the adsorption energies and the number of boron atoms, n, in the system. For boron-containing models, optimized adsorption structures are given for each system.

The situation is somewhat different when looking at the adsorption of O (Figure 2, middle). The preferred site is the bridge site, which in the case of pristine graphene implies the site between two carbon atoms. In the investigated cases of B-doped graphene, the preferred bridge site is the C-B bridge. The energy of adsorption behaves in the same way as for the H adoption—the boron percentage and adsorption energies of O are negatively correlated, the lowest energy being −4.30 eV for $O@C_{51}B_3$.

The adsorption of OH (Figure 2, bottom) gave similar results as in previously conducted research [12]—OH binds via creating a C-O bond in the case of pristine graphene and a B-O bond in the case of $C_{54-n}B_n$. When there are two boron atoms in the model, the adsorption to each one of them gives similar results, with the more stable one having a binding energy of −2.21 eV and B-O and O-H bond lengths of 1.52 Å and 0.98 Å, respectively. However, for the $C_{51}B_3$ structure, it is most favourable for the OH group to adsorb on top of the boron atom that is in the middle, obtaining a binding energy of −2.21 eV and the same bond lengths as in $C_{52}B_2$. It is also interesting to note that, unlike the cases where the adsorption of H and O was investigated, the most stable model with adsorbed OH is not the $C_{51}B_3$ structure but $C_{52}B_2$. The bond lengths do not vary much when increasing the number of boron atoms. It is also noted that the re-hybridization from sp^2 to sp^3 occurs in all the cases upon H, O or OH adsorption, meaning that the models lose their initial planarity.

Obviously, all the investigated adsorbates bind more strongly to B-doped graphene than to pristine graphene, with the strength of the interaction rising with B percentage. This means that the introduction of B into the graphene lattice results in an enhanced reactivity of the material towards H, O and OH. The increased reactivity towards H is of significance for the possible applications of such materials for hydrogen production and storage. The adsorption energy of atomic hydrogen is used as one of the basic descriptors for the hydrogen evolution reaction (HER) activity of the material. In fact, metal-free substitutionally B-doped graphene (1.85 at.% of B) was previously demonstrated to have significantly higher HER activity compared to defective graphene and glassy carbon, with a HER onset potential of around −0.3 V vs. RHE (Reversible Hydrogen Electrode) [44]. This value agrees well with the calculated hydrogen binding energy on B-doped graphene (~−2.1 eV, Figure 2). Moreover, from the results presented here, it is clear that the HER active sites are not the B site but the carbon atoms adjacent to the dopant sites, whose activity is rendered by B incorporation into the graphene lattice.

On the other hand, the enhanced reactivity of $C_{54-n}B_n$ towards O-containing groups signals that boron doping makes graphene more prone to oxidation, i.e., that O-groups will be more stable on the surface containing B than the pristine one. This is very important from the aspect of metal interaction with such surfaces, which is crucial for applications of materials in metal-ion batteries. It is known that the oxygen-containing groups on graphene can play a very important role in this case, allowing for a stronger interaction of the chosen metal with the surface [38,40]. However, there is a risk of the O-group interacting "too strongly" with the metal and detaching from the surface irreversibly. Therefore, the stabilization of O-groups on graphene by its doping with B could be a good strategy for the better performance of graphene-based materials in metal-ion batteries. Another important question, when it comes to electrochemical applications, is the electrical conductivity of the material. The electronic structures of $C_{54-n}B_n$ upon the adsorption of H, O and OH are given in Figure 3. It can be seen that the oxidized forms of $C_{54-n}B_n$ (i.e., $C_{54-n}B_n$ with O and OH adsorbed) act as conductors, making them good candidates for further research. However, we note that there is a DOS change around the Fermi level upon oxidation, which indicates that even though there is no band gap opening, the conductivity of the material is altered by the presence of oxygen-containing groups. The adsorption of the two metals of interest, Na and Al, onto bare and oxidized $C_{54-n}B_n$ surfaces will be investigated in the next sections.

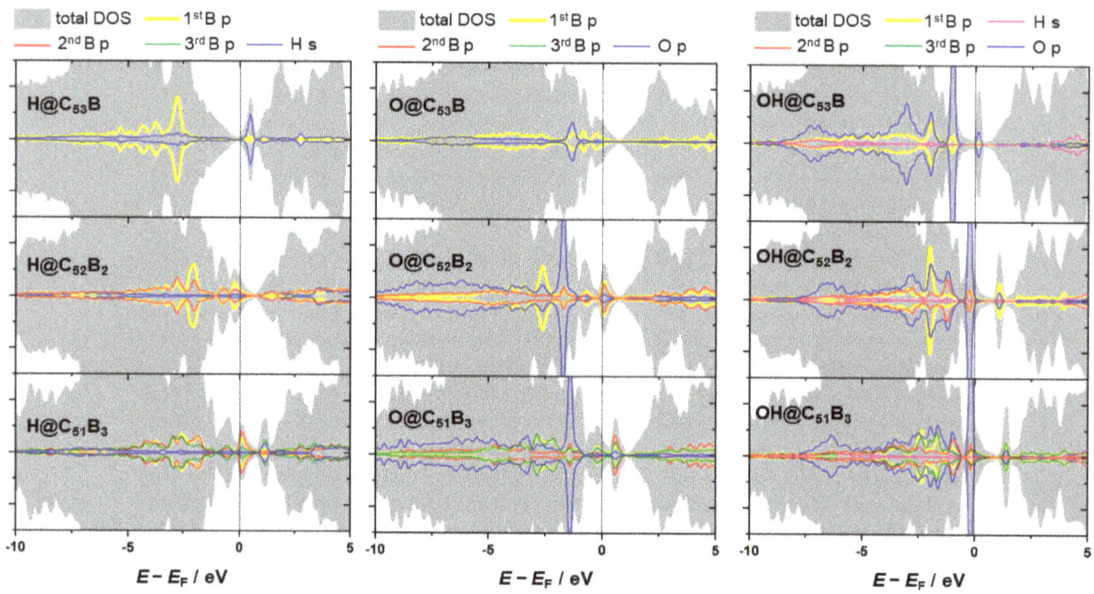

Figure 3. Electronic structures of the optimized H@ $C_{54-n}B_n$ (**left**), O@$C_{54-n}B_n$ (**middle**) and OH@$C_{54-n}B_n$ systems (**right**). Except for the total DOSs and p states of dopant atoms, the states of the adsorbate are also given as follows: p states of O (cases of O_{ads} and OH_{ads}) and s states of H (cases of H_{ads} and OH_{ads}).

3.3. Metal Adsorption

Sodium is known to adsorb on pristine graphene at the C_6-hollow site with an adsorption energy of −0.93 eV [16]. Aluminum also adsorbs to a hollow site on pristine graphene, with an adsorption energy amounting to −1.09 eV [16]. The reported energies vary depending on the computational approach (especially the chosen dispersion correction) and the size of the supercell. Regarding the adsorption of Na onto oxidized (nondoped) graphene, it is known that Na can directly interact with OH group, resulting in its detachment from the basal plane and NaOH phase separation [33]. We confirm all the mentioned facts regarding pristine graphene here and, additionally, investigate the adsorption of sodium and aluminum onto $C_{54-n}B_n$, O@$C_{54-n}B_n$ and OH@$C_{54-n}B_n$. The calculated metal adsorption energies on pristine graphene are found to be −0.85 eV and −1.26 eV for Na and Al, respectively. When it comes to OH@C_{54}, we find that both Na and Al adsorption result in metal hydroxyde phase separation. On the other hand, for O@C_{54}, there is no detachment of the O-group, and we find that the adsorption energy of Na is −1.54 eV and that of Al is −3.84 eV. In the case of $C_{54-n}B_n$, the metal can occupy various top, bridge, or hollow adsorption sites. For the oxidized surfaces, the metal generally interacts directly with the oxygen-containing group. Three different outcomes can be identified, and they are represented in Figure 4. Here we have shown only the representatives of the three main groups of adsorption geometries. All optimized structures of Na and Al adsorption on the investigated (oxidized) B-doped graphene can be found in SI, Figures S2 and S3.

The first outcome is for the cases where the metal atom is adsorbed on the same side of the basal plane as O or OH. In these cases, the metal atom interacts directly with the oxygen-containing group. Most of the investigated systems behave like this. In the second outcome, the adsorbed Al or Na is on the opposite side of the basal plane from oxygen. This was only found for $C_{51}B_3$ models with an adsorbed oxygen atom. This can be explained by the combination of a high dopant concentration and the presence of the O-group, which significantly alter the planarity and charge of the surface. Because of this, it is easier for the metal atom to approach the other side of the plane. In both cases, the adsorbed metal occupies a hollow site on the opposite side from O_{ads}. This is similar to the case of Na adsorption on corrugated doped graphene [45]. The third outcome represent the cases where the new phase emerges upon the adsorption of the metal. This happens in three cases: Na@OH@pristine graphene, Al@OH@pristine graphene and when OH is bounded on top of the $C_{53}B$ after the adsorption of Al. Therefore, these new phases represent the formation of metal hydroxyde. These models were not considered further since the formation of a new phase is not favourable for electrochemical purposes. Additionally, the charge redistribution is considered through charge difference plots. Again, for brevity, the mentioned plots are given in Figure 5 only for the chosen systems, which are representatives of the different types of interaction of the metals with the surfaces.

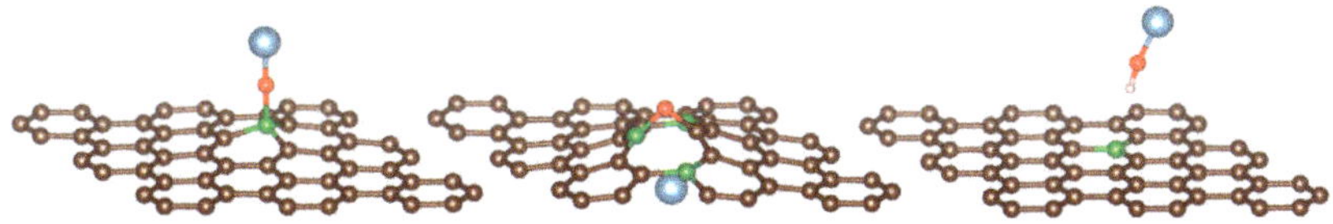

Figure 4. Examples of different outcomes of the metal adsorption on top of the obtained models for the case of Al: Al adsorbed at O@$C_{53}B$ (**left**), Al at O@$C_{51}B_3$ (**middle**) and Al at nonoxidized $C_{53}B$ (**right**).

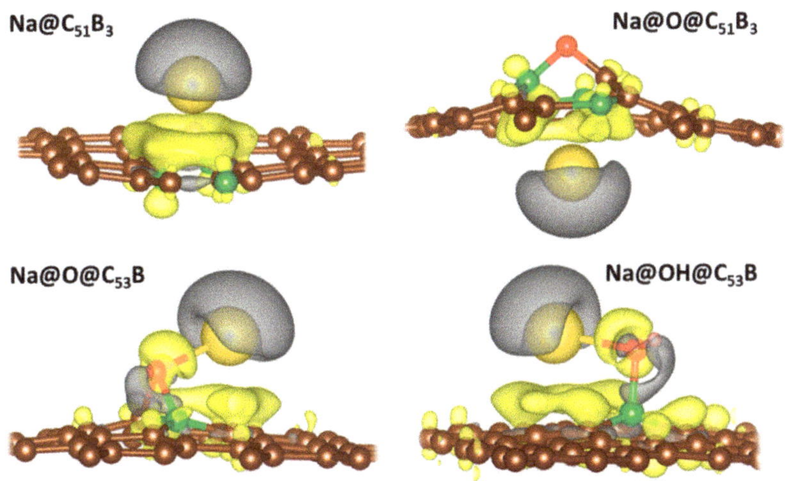

Figure 5. Charge difference plots for the indicated chosen systems of Na adsorption. Grey isosurfaces indicate charge loss, while the yellow ones indicate charge gain. The isosurface values are ±0.002 eV/Å.

It can be clearly seen that in all the cases, the metal atoms are the ones that lose their charge and transfer it either completely to the graphene basal plane (in the cases of nonoxidized surfaces) or partially to oxygen and partially to the basal plane (in the cases of oxidized surfaces). The only exception when it comes to the oxidized surfaces is the already mentioned M@O@$C_{51}B_3$ (Figure 5, top right), where M adsorbs on the opposite side of the basal plane from O, so that no charge transfer from M to O is found. According to Löwdin charges, Na transfers 0.65–0.75 e to the substrate in all the cases, while Al transfers 0.46–0.52 e to the substrate.

In order to avoid forming a metal precipitate, the interaction energy between the metal and the substrate needs to be higher than the cohesive energies of the metals [46]. Thus, one of the main points of interest was comparing the absolute value of the calculated adsorption energies with the cohesive energies of Na 1.113 eV atom^{-1}, and Al 3.39 eV atom^{-1} [47,48], as shown in Figure 6. It can be seen that in the case of sodium, only one system has an energy lower than the cohesive energy, and that is $O@C_{54}$. On the other hand, for aluminum, half of the systems, not counting the ones where a new phase emerges, do not fulfil this condition. This is intuitive since Al has higher cohesive energy compared to Na. The systems where the precipitation of Al does not occur are $O@C_{54}$, $O@C_{53}B$, $O@C_{52}B_2$, $OH@C_{52}B_2$ and $OH@C_{51}B_3$.

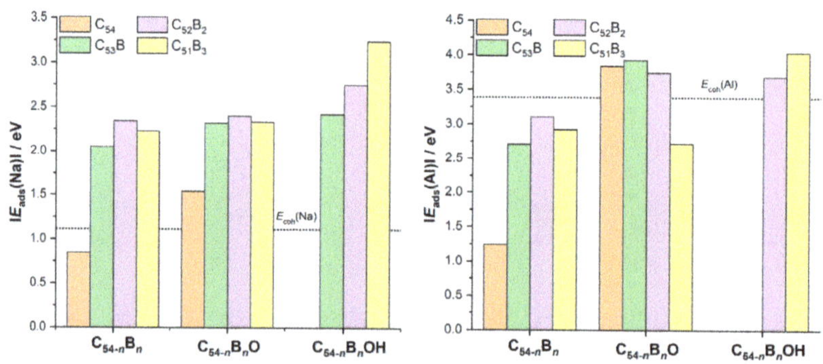

Figure 6. Adsorption energies of Na (**left**) and Al (**right**) in $C_{54-n}B_n$ models compared to their cohesive energies (E_{coh}), given by dotted lines. The missing columns ($Na@C_{54}OH$, $Al@C_{54}OH$ and $Al@C_{53}BOH$) indicate metal–hydroxyde phase separation.

For electrochemistry purposes "the intermediate" adsorption energy is the one that is preferable, meaning that it is high enough for metals not to precipitate but the adsorption is not so strong that it becomes irreversible.

For substrates without pre-adsorbed O or OH, the absolute value of the adsorption energies of sodium and aluminum increases with the increment of boron atoms in the structure up to a point where there are two boron atoms, and then it decreases again, making the $C_{52}B_2$ model the one with the most exothermic adsorption energy for both metals. The same trend is observed for Na when O is pre-adsorbed on the surface. However, this is not the case for Al, for which the highest absolute value of the adsorption energy is for the $O@C_{52}B_2$ structure. When OH is bonded to the surface, for both metals, the absolute value of the adsorption energy increases with the percentage of boron.

Another important point is determining whether the systems are conducting. They need to be conducting to be used as electrode materials. This is why the analysis of the electronic structure, in particular, the density of states (DOS), was carried out for all the investigated systems. Figure 7 shows them for the case of sodium (for brevity).

Out of the considered models that fulfil the previous conditions, almost all behave as conductors, except $Na@C_{53}B$ and $Na@O@C_{53}B$, which, upon adsorbing sodium, have a very low DOS around the Fermi energy (Figure 7).

Next, we investigated how the presence of boron dopant atoms in different concentrations, as well as the presence of O/OH groups on the surface, affect the adsorption capacity for Na and Al (Figure 8). We compare the calculated differential adsorption energies of each added Na atom to its cohesive energy and find that up to four Na atoms can be adsorbed onto $C_{51}B_3$, $O@C_{53}B$ and $O@C_{52}B_2$. For the oxidized forms of $C_{51}B_3$, the absolute value of $E_{ads,diff}$ of the fourth Na atom is lower than Na's cohesive energy, which indicates that metal phase precipitation would occur. On the other hand, for $OH@C_{53}B$, the Na-O phase separation occurs upon the addition of the second Na atom, while for $C_{53}B$ and $OH@C_{52}B_2$, it happens upon adding the third Na atom, and for $C_{52}B_2$, after the addition of the fourth Na atom. The optimized structures of two, three and four Na atom adsorption onto the models for which no phase separation occurs and the adsorption energies overcome the cohesive energy of Na are shown in Figure 9. From this figure, it is obvious that, in the case of nonoxidized boron-doped graphene, Na prefers adsorbing on B-containing hollow sites. On the other hand, in the case of the oxidized surfaces, it mostly interacts with the O-group, which can bind up to three Na atoms.

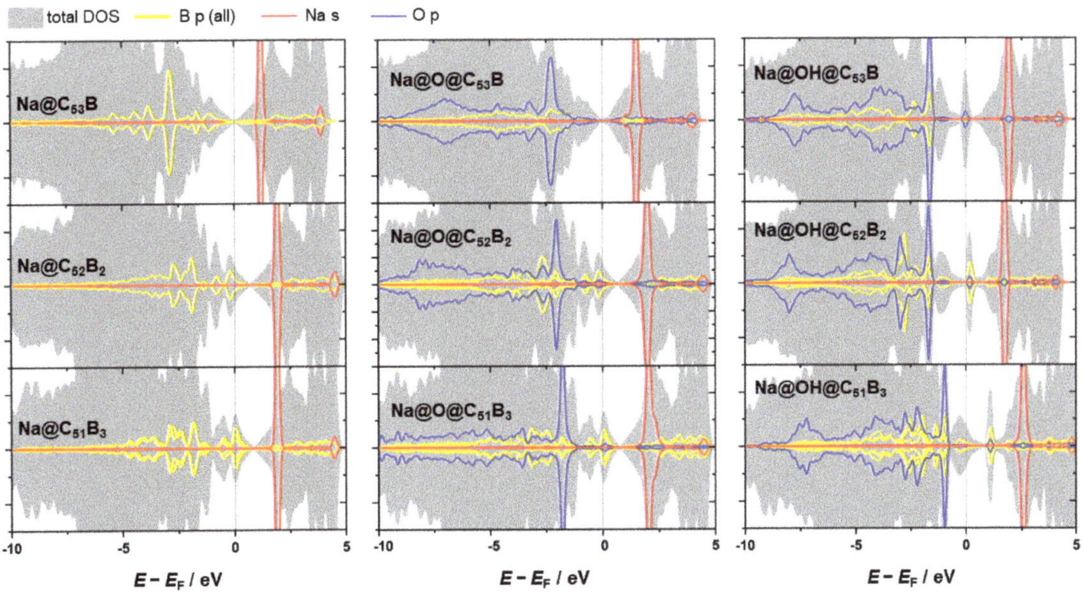

Figure 7. Electronic structures of sodium adsorption onto nonoxidized and oxidized $C_{54-n}B_n$ models. Except for total DOS, the s states of sodium are given, as well as the p states of boron dopant atoms and p states of O (when present, as in the cases of O@ $C_{54-n}B_n$ and OH@$C_{54-n}B_n$).

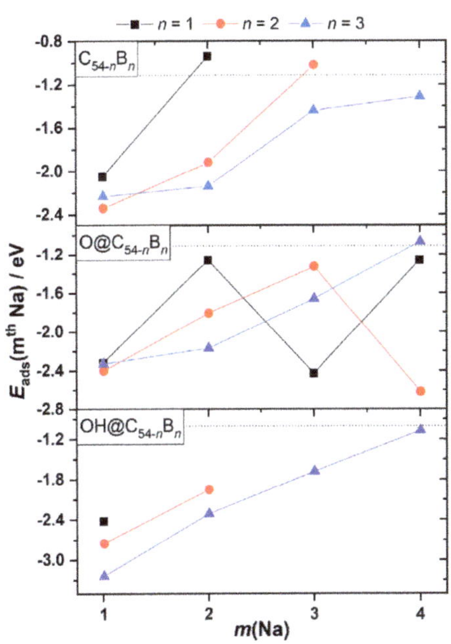

Figure 8. Differential adsorption energies of m Na atoms (m goes from 1 to 4) onto the investigated substrates. Lacking data points indicate M-O phase separation (e.g., in the case of 2Na@OH@$C_{53}B$). For comparison, the cohesive energy of Na is given as dotted, grey horizontal lines.

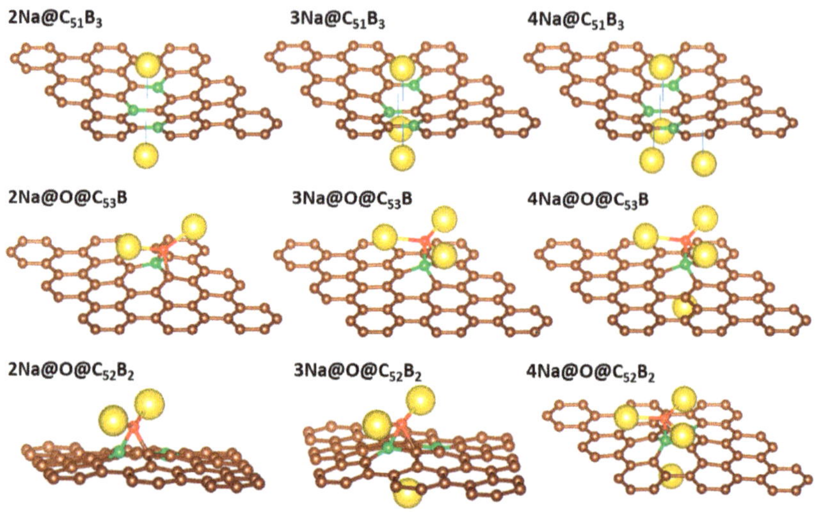

Figure 9. Optimized structures of 2–4 Na adsorption onto $C_{51}B_3$ (**top row**), O@C_{53}B (**middle row**), and O@$C_{52}B_2$ (**bottom row**).

When it comes to Al adsorption (Table 2), upon the addition of the second Al atom, most cases result in adsorption energies lower than the Al cohesive energy (comparing absolute values), while for others, Al-O phase separation occurs. The only case with a stable adsorption of 2Al is 2Al@O@$C_{52}B_2$, which also results in Al-O phase separation upon the addition of the third Al atom.

Table 2. Differential adsorption energies of 1-2 Al atoms onto the investigated substrates. Lacking values indicate M-O phase separation. Bolded values indicate that the adsorption energy overcomes Al's cohesive energy.

m(Al)	E_{ads}(m^{th} Al@subs)/eV								
	C_{53}B	$C_{52}B_2$	$C_{51}B_3$	O@ C_{53}B	O@ $C_{52}B_2$	O@ $C_{51}B_3$	OH@ C_{53}B	OH@ $C_{52}B_2$	OH@ $C_{51}B_3$
1	−2.71	−3.11	−2.92	**−3.93**	**−3.75**	−2.72	**−3.80**	**−3.69**	**−4.05**
2	-	-	-	-	**−3.70**	-	-	-	−2.81

Finally, as we see clear improvement in the materials' capacity for Na induced by B-doping, we calculate the sodiation potentials of the studied materials as electrode materials in a galvanic cell. For this consideration, a proper reference point needs to be taken, since potential cannot be directly compared to the adsorption energies of an isolated Na atom. Therefore, we use the approach explained in detail in ref. [33] and consider a galvanic cell (potential secondary sodium-ion battery) with solid Na as the anode and the investigated boron-doped graphene surfaces as the cathodes. The half reactions in this cell (equivalent to battery discharging) would be, at the anode:

$$\mathrm{Na}_{(s)} \rightarrow \mathrm{Na}^+ + \mathrm{e}^- \tag{6}$$

and at the cathode:

$$\mathrm{Na}^+ + \mathrm{e}^- + {}^*(\mathrm{O/OH})\mathrm{C}_{54-n}\mathrm{B}_n \rightarrow \mathrm{Na@(O/OH)C}_{54-n}\mathrm{B}_n, \tag{7}$$

where *(O/OH)$C_{54-n}B_n$ stands for the adsorption site on the investigated surface, i.e., $C_{54-n}B_n$, O@$C_{54-n}B_n$ or OH@$C_{54-n}B_n$. We note that this consideration assumes that the activity of Na^+ in the electrolyte is 1, that the changes in volume and entropy due to adsorption are negligible [49], and that the Na^+ ions interact with B-graphene free of their solvation spheres. This way, the change in the Gibbs free energy of the adsorption, ΔG, can be estimated as the energy change in the Na transfer from the metallic Na anode to the B-graphene cathode. As explained in ref. [33], under the mentioned assumptions, the electrode potential at which the sodiation of graphene-based material takes place, E^0 (Na^+/Na@(O/OH)$C_{54-n}B_n$), referred to as the Standard Hydrogen Electrode, SHE, is

$$E^0(\mathrm{Na^+/Na@(O/OH)C_{54-\mathit{n}}B_\mathit{n}}) \approx E^0(\mathrm{Na^+/Na}) - \Delta E/F \tag{8}$$

where E^0(Na^+/Na) is the standard electrode potential of the Na^+/Na electrode (−2.71 V vs. SHE, from tables of standard electrode potentials), ΔE is the energy change upon Na adsorption and F is the Faraday constant. When we look at the average adsorption of 1Na (1e^- in the reaction), ΔE reduces to $E_{ads,int}$(Na) + E_{coh}(Na), and we get

$$E^0(\mathrm{Na^+/Na@(O/OH)C_{54-\mathit{n}}B_\mathit{n}}) \approx E^0(\mathrm{Na^+/Na}) - [E_{ads}(\mathrm{Na}) + E_{coh}(\mathrm{Na})]/e \tag{9}$$

where E_{coh}(Na) is the already mentioned cohesive energy of Na, and $E_{ads,int}$(Na) is the integral (average) Na adsorption energy (both E_{coh} and $E_{ads,int}$ are given in eVs, per one Na atom, and therefore divided only by e in Equation (9), where E_{coh}(Na) = 1.113 eV $atom^{-1}$ [47,48]). From Equation (9), it can be seen that E^0(Na^+/Na@(O/OH)$C_{54-n}B_n$) is a linear function of E_{ads}(Na). Given that E_{ads}(Na) is negative, the stronger the Na binding is to the surface, the more positive E^0(Na^+/Na@(O/OH)$C_{54-n}B_n$) will be. The calculated sodiation potentials are shown in Table 3.

Table 3. Integral adsorption energies ($E_{ads,int}$) of 2–4 Na atoms on (O/OH)$C_{54-n}B_n$ surfaces and corresponding sodiation potentials (E(Na^+/Na@subs, bold values), where subs = (O/OH)$C_{54-n}B_n$). "*" indicates that phase separation has occurred, and such systems were not taken into further account.

Subs	2Na		3Na		4Na	
	$E_{ads,int}$ /eV	E^0(Na^+/Na@subs) /V	$E_{ads,int}$ /eV	E^0(Na^+/Na@subs) /V	$E_{ads,int}$ /eV	E^0(Na^+/Na@subs) /V
$C_{53}B$	−1.50	**−2.34**	*	*	*	*
$C_{52}B_2$	−2.13	**−1.71**	*	*	*	*
$C_{51}B_3$	−2.19	**−1.65**	−1.94	**−1.90**	−1.78	**−2.06**
O@$C_{53}B$	−1.79	**−2.05**	−2.00	**−1.84**	−1.82	**−2.02**
O@$C_{52}B_2$	−2.10	**−1.74**	−1.84	**−2.00**	−2.04	**−1.80**
O@$C_{51}B_3$	−2.25	**−1.59**	−2.05	**−1.79**	*	*
OH@$C_{52}B_2$	−2.35	**−1.49**	*	*	*	*
OH@$C_{51}B_3$	−2.77	**−1.07**	−2.41	**−1.43**	*	*

It is obvious from Table 3 that the calculated sodiation potentials for all the investigated (oxidized) B-doped graphene surfaces are negative, with −2.34 V vs. SHE for Na^+/Na@$C_{53}B$ for m = 2 as the lowest (most negative). However, none of them are more negative than the standard electrode potential of the Na^+/Na electrode (−2.71 V vs. SHE), which indicates that in the considered galvanic cell, solid Na would indeed dissolve to release Na^+ ions, which would then adsorb at the B-doped surfaces of graphene. The material with the most positive sodiation potential (the case of Na^+/Na@OH@$C_{51}B_3$ for m = 2) combined with an Na^+/Na electrode would yield the most positive standard electromotive force (ε^0) of the cell, ε^0 = −1.07 V − (−2.71 V) = 1.64 V, resulting in a high-voltage sodium-ion battery. The possibility of obtaining high-voltage sodium-ion batteries is in line with our previous findings on (oxidized) doped graphene, ref. [33]. However, in the mentioned study, the dopant concentration was not investigated as a variable. Here, we can clearly see (Table 3) that for the case of 2Na adsorption, an increase in B concentration makes the sodiation potential more positive. However, this is not the case when we consider three and four Na atom adsorption, where $C_{52}B_2$ is found to be optimal.

It is interesting to compare pristine graphene and its oxidized forms with the boron-doped analogues studied here. If we were to calculate the sodiation potential for pristine, nonoxidized graphene, based only on 1Na adsorption, it renders a very negative $E^0(Na^+/Na@C_{54})$, amounting to −2.99 V. Obviously, this case surpasses (is more negative than) the standard electrode potential for pure Na^+/Na, indicating that in such a galvanic cell, Na^+/Na would actually act as the cathode, where Na^+ would reduce, while $Na^+/Na@C_{54}$ would act as the anode, i.e., the half reactions (R1) and (R2) would go in opposite directions than previously stated. The resulting ε^0 would be 0.28 V. This means that the pristine graphene material would have to be filled with Na_{ads} before the start of the battery operation to provide Na, which would oxidate during the battery discharge. Additionally, next to low battery voltage, we add the fact that pristine graphene shows a rather poor Na capacity, with already the second Na not being able to surpass $E_{coh}(Na)$, making room for Na–metal phase separation and deteriorating the possibility of preparing the Na-filled graphene electrode material. When it comes to the oxidized forms of non-doped graphene, in O-functionalized graphene, one epoxy group is found to be able to bind up to three Na atoms, while the presence of OH allows for its interaction with just one Na. The B-doping of the material is found to improve this situation—the sodiation potentials are more positive than Na^+/Na, the cohesive energy problem is surpassed and the Na^+ (battery) capacity is improved up to four times.

4. Conclusions

DFT calculations were utilized to investigate the influence of boron dopant concentration and surface oxidation on the potential for Na and Al storage in graphene. The obtained results demonstrated that the introduction of boron into graphene has a significant effect on its reactivity towards oxygen-containing species, rendering the material more susceptible to oxidation. Consequently, when modeling such materials, it is crucial to consider the oxidation of the B centers. In terms of electrochemical applications, the oxidation of B-doped graphene modifies its electronic structure but does not induce a bandgap. As a result, the controllable doping of graphene with boron, followed by oxidation, appears to be a promising approach for enhancing its electrochemical performance with respect to Na and Al storage. Specifically, this strategy optimizes the strength of the interaction between Na and Al and the graphene surface and its storage capacity, thereby opening up possibilities for its applications in metal-ion batteries and energy storage in general. In both cases, the adsorbate atoms tend to bind to or near B atoms, which are additionally functionalized by surface groups such as O or OH. Depending on the B-dopant concentration and the oxidation of the surface, such functionalized materials can store up to four Na atoms per dopant atom and are possible candidates for high-voltage (up to approx. 1.6 V) and high-capacity Na-ion batteries.

Supplementary Materials: The following supporting information can be downloaded at https://www.mdpi.com/article/10.3390/c9040092/s1: Figure S1. Investigated configurations of $C_{52}B_2$. Figure S2. Investigated configurations of $C_{51}B_3$. Figure S3. Electronic structures (total DOS) of pristine graphene (C_{54}, first row) and investigated boron-doped graphene $C_{54-n}B_n$ (bottom three rows), zoomed in to the range of −5 to +5 eV (left) and in full energy range (right). The p states of boron dopant atoms are shown separately (in the case of $C_{52}B_2$, the p states of two B atoms overlap). For clarity, all total DOS plots are divided by 5. The Fermi level (dashed, black line) is set to 0. Figure S4. Optimized structures of Na adsorption onto (oxidized) $C_{54-n}B_n$ systems (first row: bare $C_{54-n}B_n$, middle row: $C_{54-n}B_n$ oxidized by O_{ads}, bottom row: $C_{54-n}B_n$ oxidized by OH_{ads}). Graphical representations were made using VESTA. Figure S5. Optimized structures of Al adsorption onto (oxidized) $C_{54-n}B_n$ systems (first row: bare $C_{54-n}B_n$, middle row: $C_{54-n}B_n$ oxidized by O_{ads}, bottom row: $C_{54-n}B_n$ oxidized by OH_{ads}). Graphical representations were made using VESTA.

Author Contributions: Conceptualization, A.S.D. and I.A.P.; data curation, N.V.S. and I.A.P.; formal analysis, M.S.R. and A.S.D.; funding acquisition, N.V.S. and I.A.P.; investigation, M.S.R. and A.S.D.; methodology, M.S.R., N.V.S., A.S.D. and I.A.P.; resources, N.V.S. and I.A.P.; supervision, N.V.S. and A.S.D.; validation, N.V.S. and I.A.P.; writing—original draft, M.S.R. and A.S.D.; writing—review and editing, N.V.S. and I.A.P. All authors have read and agreed to the published version of the manuscript.

Funding: This research was supported by the Serbian Ministry of Science, Technological Development, and Innovations (contract number: 451-03-47/2023-01/200146). I.A.P. is also indebted to the Research Fund of the Serbian Academy of Sciences and Arts, project F-190, for supporting this study. The computations and data handling were enabled by resources provided by the Swedish National Infrastructure for Computing (SNIC) at the National Supercomputer Centre (NSC) at Linköping University, partially funded by the Swedish Research Council through grant agreement No. 2018-05973.

Data Availability Statement: The raw/processed data required to reproduce these findings cannot be shared at this time as the data also form part of an ongoing study.

Conflicts of Interest: The authors declare no conflict of interest.

References

1. Novoselov, K.S.; Geim, A.K.; Morozov, S.V.; Jiang, D.; Katsnelson, M.I.; Grigorieva, I.V.; Dubonos, S.V.; Firsov, A.A. Two-Dimensional Gas of Massless Dirac Fermions in Graphene. *Nature* **2005**, *438*, 197–200. [CrossRef] [PubMed]
2. Liu, J.; Liang, T.; Tu, R.; Lai, W.; Liu, Y. Redistribution of π and σ Electrons in Boron-Doped Graphene from DFT Investigation. *Appl. Surf. Sci.* **2019**, *481*, 344–352. [CrossRef]
3. Brownson, D.A.C.; Kampouris, D.K.; Banks, C.E. An Overview of Graphene in Energy Production and Storage Applications. *J. Power Sources* **2011**, *196*, 4873–4885. [CrossRef]
4. Shanmugam, S.; Nachimuthu, S.; Subramaniam, V. DFT Study of Adsorption of Ions on Doped and Defective Graphene. *Mater. Today Commun.* **2020**, *22*, 100714. [CrossRef]
5. Ali, S.; Lone, B. Adsorption of Cytosine on Si and Ge Doped Graphene: A DFT Study. *Mater. Today Proc.* **2023**, *80*, 774–781. [CrossRef]
6. Tiwari, S.K.; Sahoo, S.; Wang, N.; Huczko, A. Graphene Research and Their Outputs: Status and Prospect. *J. Sci. Adv. Mater. Devices* **2020**, *5*, 10–29. [CrossRef]
7. Soldano, C.; Mahmood, A.; Dujardin, E. Production, Properties and Potential of Graphene. *Carbon* **2010**, *48*, 2127–2150. [CrossRef]
8. Agnoli, S.; Favaro, M. Doping Graphene with Boron: A Review of Synthesis Methods, Physicochemical Characterization, and Emerging Applications. *J. Mater. Chem. A* **2016**, *4*, 5002–5025. [CrossRef]
9. Liu, L.; Qing, M.; Wang, Y.; Chen, S. Defects in Graphene: Generation, Healing, and Their Effects on the Properties of Graphene: A Review. *J. Mater. Sci. Technol.* **2015**, *31*, 599–606. [CrossRef]
10. Yang, G.; Li, L.; Lee, W.B.; Ng, M.C. Structure of Graphene and Its Disorders: A Review. *Sci. Technol. Adv. Mater.* **2018**, *19*, 613–648. [CrossRef]
11. Riyaz, M.; Garg, S.; Kaur, N.; Goel, N. Boron Doped Graphene as Anode Material for Mg Ion Battery: A DFT Study. *Comput. Theor. Chem.* **2022**, *1214*, 113757. [CrossRef]
12. Dobrota, A.S.; Pašti, I.A.; Mentus, S.V.; Skorodumova, N.V. A DFT Study of the Interplay between Dopants and Oxygen Functional Groups over the Graphene Basal Plane-Implications in Energy-Related Applications. *Phys. Chem. Chem. Phys.* **2017**, *19*, 8530–8540. [CrossRef] [PubMed]
13. Zhang, Q.; Zhou, Y.; Yu, Y.; Chen, B.-Y.; Hong, J. Exploring Catalytic Performance of Boron-Doped Graphene Electrode for Electrochemical Degradation of Acetaminophen. *Appl. Surf. Sci.* **2020**, *508*, 145111. [CrossRef]
14. Etacheri, V.; Marom, R.; Elazari, R.; Salitra, G.; Aurbach, D. Challenges in the Development of Advanced Li-Ion Batteries: A Review. *Energy Environ. Sci.* **2011**, *4*, 3243–3262. [CrossRef]
15. Liang, Y.; Dong, H.; Aurbach, D.; Yao, Y. Current Status and Future Directions of Multivalent Metal-Ion Batteries. *Nat. Energy* **2020**, *5*, 646–656. [CrossRef]
16. Pašti, I.A.; Jovanović, A.; Dobrota, A.S.; Mentus, S.V.; Johansson, B.; Skorodumova, N.V. Atomic Adsorption on Pristine Graphene along the Periodic Table of Elements—From PBE to Non-Local Functionals. *Appl. Surf. Sci.* **2018**, *436*, 433–440. [CrossRef]
17. Lin, M.C.; Gong, M.; Lu, B.; Wu, Y.; Wang, D.Y.; Guan, M.; Angell, M.; Chen, C.; Yang, J.; Hwang, B.J.; et al. An Ultrafast Rechargeable Aluminium-Ion Battery. *Nature* **2015**, *520*, 324–328. [CrossRef]
18. Zhou, W.; Jia, J.; Lu, J.; Yang, L.; Hou, D.; Li, G.; Chen, S. Recent Developments of Carbon-Based Electrocatalysts for Hydrogen Evolution Reaction. *Nano Energy* **2016**, *28*, 29–43. [CrossRef]

19. Zhang, T.; Li, C.; Wang, F.; Noori, A.; Mousavi, M.F.; Xia, X.; Zhang, Y. Recent Advances in Carbon Anodes for Sodium-Ion Batteries. *Chem. Rec.* **2022**, *22*, e202200083. [CrossRef]
20. Benti, N.E.; Tiruye, G.A.; Mekonnen, Y.S. Boron and Pyridinic Nitrogen-Doped Graphene as Potential Catalysts for Rechargeable Non-Aqueous Sodium–Air Batteries. *RSC Adv.* **2020**, *10*, 21387–21398. [CrossRef]
21. Panchakarla, L.S.; Subrahmanyam, K.S.; Saha, S.K.; Govindaraj, A.; Krishnamurthy, H.R.; Waghmare, U.V.; Rao, C.N.R. Synthesis, Structure, and Properties of Boron- and Nitrogen-Doped Graphene. *Adv. Mater.* **2009**, *21*, 4726–4730. [CrossRef]
22. Yu, X.; Han, P.; Wei, Z.; Huang, L.; Gu, Z.; Peng, S.; Ma, J.; Zheng, G. Boron-Doped Graphene for Electrocatalytic N_2 Reduction *Joule* **2018**, *2*, 1610–1622. [CrossRef]
23. Rao, C.N.R.; Gopalakrishnan, K.; Govindaraj, A. Synthesis, Properties and Applications of Graphene Doped with Boron, Nitrogen and Other Elements. *Nano Today* **2014**, *9*, 324–343. [CrossRef]
24. Liu, Y.; Artyukhov, V.I.; Liu, M.; Harutyunyan, A.R.; Yakobson, B.I. Feasibility of Lithium Storage on Graphene and Its Derivatives. *J. Phys. Chem. Lett.* **2013**, *4*, 1737–1742. [CrossRef]
25. Hao, Q.; Xia, X.; Lei, W.; Wang, W.; Qiu, J. Facile Synthesis of Sandwich-like Polyaniline/Boron-Doped Graphene Nano Hybrid for Supercapacitors. *Carbon* **2015**, *81*, 552–563. [CrossRef]
26. Zhang, L.; Zhang, Z.Y.; Liang, R.P.; Li, Y.H.; Qiu, J.D. Boron-Doped Graphene Quantum Dots for Selective Glucose Sensing Based on the "Abnormal" Aggregation-Induced Photoluminescence Enhancement. *Anal. Chem.* **2014**, *86*, 4423–4430. [CrossRef]
27. Choudhuri, I.; Patra, N.; Mahata, A.; Ahuja, R.; Pathak, B. B-N@Graphene: Highly Sensitive and Selective Gas Sensor. *J. Phys. Chem. C* **2015**, *119*, 24827–24836. [CrossRef]
28. Fan, Z.; Li, Y.; Li, X.; Fan, L.; Zhou, S.; Fang, D.; Yang, S. Surrounding Media Sensitive Photoluminescence of Boron-Doped Graphene Quantum Dots for Highly Fluorescent Dyed Crystals, Chemical Sensing and Bioimaging. *Carbon* **2014**, *70*, 149–156. [CrossRef]
29. Ling, C.; Mizuno, F. Boron-Doped Graphene as a Promising Anode for Na-Ion Batteries. *Phys. Chem. Chem. Phys.* **2014**, *16*, 10419–10424. [CrossRef]
30. Zhang, Y.; Qin, J.; Batmunkh, M.; Zhong, Y.L. Facile Synthesis of Boron-Doped Reduced Electrochemical Graphene Oxide for Sodium Ion Battery Anode. *JOM* **2021**, *73*, 2531–2539. [CrossRef]
31. Du, Y.; Zhang, B.; Kang, R.; Zhou, W.; Zhang, W.; Jin, H.; Wan, J.; Zhang, J.; Chen, G. Boron-Doping-Induced Defect Engineering Enables High Performance of a Graphene Cathode for Aluminum Batteries. *Inorg. Chem. Front.* **2022**, *9*, 925–934. [CrossRef]
32. Diklić, N.P.; Dobrota, A.S.; Pašti, I.A.; Mentus, S.V.; Johansson, B.; Skorodumova, N.V. Sodium Storage via Single Epoxy Group on Graphene—The Role of Surface Doping. *Electrochim. Acta* **2019**, *297*, 523–528. [CrossRef]
33. Dobrota, A.S.; Pašti, I.A.; Mentus, S.V.; Johansson, B.; Skorodumova, N.V. Functionalized Graphene for Sodium Battery Applications: The DFT Insights. *Electrochim. Acta* **2017**, *250*, 185–195. [CrossRef]
34. Giannozzi, P.; Baroni, S.; Bonini, N.; Calandra, M.; Car, R.; Cavazzoni, C.; Ceresoli, D.; Chiarotti, G.L.; Cococcioni, M.; Dabo, I.; et al. QUANTUM ESPRESSO: A Modular and Open-Source Software Project for Quantum Simulations of Materials. *J. Phys. Condens. Matter* **2009**, *21*, 395502. [CrossRef] [PubMed]
35. Giannozzi, P.; Andreussi, O.; Brumme, T.; Bunau, O.; Buongiorno Nardelli, M.; Calandra, M.; Car, R.; Cavazzoni, C.; Ceresoli, D.; Cococcioni, M.; et al. Advanced Capabilities for Materials Modelling with Quantum ESPRESSO. *J. Phys. Condens. Matter* **2017**, *29*, 465901. [CrossRef] [PubMed]
36. Grimme, S. Semiempirical GGA-Type Density Functional Constructed with a Long-Range Dispersion Correction. *J. Comput. Chem.* **2006**, *27*, 1787–1799. [CrossRef] [PubMed]
37. Monkhorst, H.J.; Pack, J.D. Special Points for Brillouin-Zone Integrations. *Phys. Rev. B* **1976**, *13*, 5188–5192. [CrossRef]
38. Momma, K.; Izumi, F. VESTA 3 for Three-Dimensional Visualization of Crystal, Volumetric and Morphology Data. *J. Appl. Crystallogr.* **2011**, *44*, 1272–1276. [CrossRef]
39. Dobrota, A.S.; Pašti, I.A.; Skorodumova, N.V. Oxidized Graphene as an Electrode Material for Rechargeable Metal-Ion Batteries—A DFT Point of View. *Electrochim. Acta* **2015**, *176*, 1092–1099. [CrossRef]
40. Dobrota, A.S.; Gutić, S.; Kalijadis, A.; Baljozović, M.; Mentus, S.V.; Skorodumova, N.V.; Pašti, I.A. Stabilization of Alkali Metal Ions Interaction with OH-Functionalized Graphene: Via Clustering of OH Groups-Implications in Charge Storage Applications. *RSC Adv.* **2016**, *6*, 57910–57919. [CrossRef]
41. Cuxart, M.G.; Perilli, D.; Tömekce, S.; Deyerling, J.; Haag, F.; Muntwiler, M.; Allegretti, F.; Di Valentin, C.; Auwärter, W. Spatial Segregation of Substitutional B Atoms in Graphene Patterned by the Moiré Superlattice on Ir(111). *Carbon* **2023**, *201*, 881–890. [CrossRef]
42. Pašti, I.A.; Jovanović, A.; Dobrota, A.S.; Mentus, S.V.; Johansson, B.; Skorodumova, N.V. Atomic Adsorption on Graphene with a Single Vacancy: Systematic DFT Study through the Periodic Table of Elements. *Phys. Chem. Chem. Phys.* **2018**, *20*, 858–865. [CrossRef] [PubMed]
43. Miwa, R.H.; Martins, T.B.; Fazzio, A. Hydrogen Adsorption on Boron Doped Graphene: An Ab Initio Study. *Nanotechnology* **2008**, *19*, 155708. [CrossRef] [PubMed]
44. Sathe, B.R.; Zou, X.; Asefa, T. Metal-Free B-Doped Graphene with Efficient Electrocatalytic Activity for Hydrogen Evolution Reaction. *Catal. Sci. Technol.* **2014**, *4*, 2023–2030. [CrossRef]
45. Dobrota, A.S.; Pašti, I.A.; Mentus, S.V.; Johansson, B.; Skorodumova, N.V. Altering the Reactivity of Pristine, N- and P-Doped Graphene by Strain Engineering: A DFT View on Energy Related Aspects. *Appl. Surf. Sci.* **2020**, *514*, 145937. [CrossRef]

46. Oh, J.S.; Kim, K.N.; Yeom, G.Y. Graphene Doping Methods and Device Applications. *J. Nanosci. Nanotechnol.* **2014**, *14*, 1120–1133. [CrossRef]
47. Kittel, C. *Introduction to Solid State Physics*, 8th ed.; Wiley & Sons: Hoboken, NJ, USA, 2004.
48. Kaxiras, E. *Atomic and Electronic Structure of Solids*; Cambridge University Press: Cambridge, UK, 2010.
49. Yao, L.H.; Cao, M.S.; Yang, H.J.; Liu, X.J.; Fang, X.Y.; Yuan, J. Adsorption of Na on Intrinsic, B-Doped, N-Doped and Vacancy Graphenes: A First-Principles Study. *Comput. Mater. Sci.* **2014**, *85*, 179–185. [CrossRef]

C Journal of Carbon Research

MDPI

Article

Highly Effective Electrochemical Water Splitting with Enhanced Electron Transfer between Ni_2Co Layered Double Hydroxide Nanosheets Dispersed on Carbon Substrate

Ziyi Wan [1], Ping Tang [1], Luwei Dai [1], Yao Yang [1], Lu Li [2,3], Jun Liu [4], Min Yang [1,*] and Guowei Deng [1,*]

[1] Sichuan Provincial Key Laboratory for Structural Optimization and Application of Functional Molecules, College of Chemistry and Life Science, Chengdu Normal University, Chengdu 611130, China; m19950528261@163.com (L.D.)

[2] Key Laboratory of Auxiliary Chemistry and Technology for Chemical Industry, Ministry of Education, Shaanxi University of Science and Technology, Xi'an 710021, China

[3] Shaanxi Collaborative Innovation Center of Industrial Auxiliary Chemistry and Technology, Shaanxi University of Science and Technology, Xi'an 710021, China

[4] Medical Imaging Key Laboratory of Sichuan Province, North Sichuan Medical College, Nanchong 637000, China; ustbliujun@iccas.ac.cn

[*] Correspondence: yangmin820525@126.com (M.Y.); guoweideng86@163.com (G.D.)

Abstract: A reasonable design of nickel-based catalysts is key to efficient and sustainable energy conversion. For electrocatalytic materials in alkaline electrolytes, however, atomic-level control of the active sites is essential. Moreover, the well-defined surface structure contributes to a deeper understanding of the catalytic mechanism. Here, we report the loading of defective nickel–cobalt layered double hydroxide nanosheets (Ni_2Co-LDH@C) after carbonization of silk. Under the precise regulation of the local coordination environment of the catalytic active site and the presence of defects, Ni_2Co-LDH@C can provide an ultra-low overpotential of 164.8 mV for hydrogen evolution reactions (HERs) at 10 mA cm^{-2}, exceeding that of commercial Pt/C catalysts. Density functional theory calculations show that Ni_2Co-LDH@C optimizes the adsorption energy of the intermediate and promotes the O-O coupling of the active site in the oxygen evolution reaction. When using Ni_2Co-LDH@Cs as cathodes and anodes to achieve overall water splitting, a low voltage of 1.63 V is required to achieve a current density of 10 mA cm^{-2}. As an ideal model, Ni_2Co-LDH@C has excellent water splitting properties and has the potential to develop water–alkali electrocatalysts.

Keywords: electrocatalyst; synergistic effect; carbonization of silk; overall water splitting

1. Introduction

The development and search for modern, clean, economic, and sustainable energy technologies is an essential issue that needs to be addressed [1,2]. Hydrogen has a clear advantage as the most efficient and clean energy source of future concern [3–6]. Among the numerous hydrogen carriers and hydrogen production reactions, electrolytic water is considered to be a promising method for producing hydrogen with elevated purity due to its high efficiency and low cost [7–10]. Thus, HER catalysts with excellent performance and stability have been widely studied and developed. Precious metal catalysts such as Pt-based catalysts have shown excellent HER properties, but the rarity and high price of precious metals have limited their further development [11,12]. As a result, the search for inexpensive and efficient electrocatalysts that can replace precious metals has become a hot topic. The nanoflower-like N-dope carbon-encapsulated CoNiPt alloy structure with composition segregation exhibits excellent HER activity in alkaline media [13]. However, due to their conjugated electronic structure, graphene carbon atoms are not excellent electrocatalytic active sites and exhibit poor electrocatalytic activity in HER. Silk has a naturally occurring strong molecular fiber that contains C, N, O, H, and additional elements.

Compared to conventional substrate materials, elevated-temperature carbonized silk has better electrical conductivity, richer surface structure, higher stability and flexibility, and the self-supporting structural system gives a richer means of modification [14,15].

Carbon carriers are widely used in the field of electrocatalysis [16]. However, the inert surface of carbon materials usually causes the interaction force between the metal nanoparticles and the carbon surface to be too weak, which means that it is difficult to effectively regulate the electronic structure of metal nanoparticles and improve their catalytic activity, and it is difficult to effectively inhibit their agglomeration [17,18]. To solve the above key problems, a strategy of adjusting the π-conjugated structure of silk after carbonization to enhance the interaction between the carbon carrier and the loaded metal nanoparticles was proposed to achieve effective regulation of the electronic structure and catalytic activity of metal nanoparticles, and to improve their catalytic stability. [19] Composite nanoreactors with transition metal atoms and carbon substrates have been successfully constructed by doping carbonized silk with transition metal atoms such as Co and Ni [20]. The first-principle calculation shows that the modification of metal Co and Ni atoms can lead to the redistribution of the surface charge of carbon materials, and make carbon atoms around Co and Ni atoms appear to be electron-deficient, which significantly enhances the electron transfer from Co and Ni nanoparticles to carbon carriers [21]. The effect of interfacial charge rearrangement of C nanomaterials induced by metal Co and Ni atoms doping on HER catalytic performance was investigated. Overall, the adsorption of hydrogen by Co is quite strong, resulting in its low electrocatalytic decomposition activity for aquatic hydrogen. In contrast, the Ni site at the interface of the composite nanoreactor has a moderate hydrogen adsorption capacity and elevated hydrogen evolution activity. The Ni sites further away from the interface favor the splitting of water molecules and provide an efficient source of oxygen evolution reactions (OER) [22]. As a result, carbon materials were employed as a substrate for the synthesis of various Ni-Co electrocatalysts with different ratios. Among these catalysts, Ni_2Co-LDH@C exhibited exceptional HER performance in the alkaline. DFT calculations revealed that an increased amount of Ni facilitates the hydrolysis process during alkaline HER. It is noteworthy that the catalytic activity and surface oxygen content demonstrate a distribution pattern resembling a volcano shape, primarily attributed to Ni_{3d}-O_{2p} hybridization. The synergistic effect resulting from these factors enables the Ni_2Co-LDH@C catalyst to achieve an impressively low overpotential of 164.8/183.4 mV for HER/OER at a current density of 10 mA cm^{-2}, surpassing most previously reported catalysts in terms of performance advantages. In addition, when Ni_2Co-LDH@C is used as the cathode and anode for overall water splitting, it requires only a very low voltage of 1.82 V to for 50 mA cm^{-2}. Most importantly, its exceptional long-term stability in an alkaline environment gives it a possibility for actual use in overall water splitting.

2. Materials and Methods

2.1. Degumming of Silk Worm Cocoons

Due to the poor stability of sericin in the outer layer of the original silk, the silk needs to be degummed before use to obtain silk fibroin fiber. A certain amount of cocoons was weighed, they were split, the pupae was discard, and the cocoons were cut properly. The degummed silk was placed in 5 g/L Na_2CO_3 solution of 100:1, and the temperature of water bath was maintained at about 95 °C; after 30 min, the silk was removed. After repeated cleaning with distilled water, the silk was placed in an oven at 40 °C to dry for later use.

2.2. Synthesis of Ni_xCo_y-LDH@C

A certain quality of 17 mmol urea, 1 g PVP, $Co(NO_3)_2 \cdot H_2O$, and $Ni(NO_3)_2 \cdot 6H_2O$ were added into the beaker. Measures of 70 mL anhydrous ethanol and 10 mL deionized water were added to mix the raw materials well, then 0.5 g of degummed silk was added to the flask and refluxed in an oil bath at 90 °C for 10 h. After the solution was cooled to room temperature, the final products were collected by centrifugation, washed with deionized water:ethanol = 1:1, and dried at 40 °C for 12 h. The dried products were annealed at 800 °C for 2 h at a heating rate of 2 °C min^{-1}. After calcination, 20 mg of the product was dispersed into 10 mL deionized water, and 10 mL 1.0 M HNO_3 solution was added drop by drop, then 10 mL 1.0 M $NH_3{\cdot}H_2O$ solution was added drop by drop. After ultrasonic treatment for 10 min, centrifuge washing was performed until the colution was neutral; then the samples were dried at 60 °C for 12 h to obtain the Ni_xCo_y LDH@C.

2.3. Structure Characterizations

A D/MAX2400 X-ray diffractometer produced by Company in Japan was adopted for X-ray diffraction (XRD) analysis; CuKa (X = 1.54 Å) was used as the radiation source. The operating voltage was 200 kV, current was 50 mA, 2θ was 10–80°, and the scanning rate was 10 min^{-1}. Raman spectroscopy analysis was used (Thermo Fisher Company, Waltham, MA, USA), in which the laser wavelength was 532 nm and the detection range was 0–2500 cm^{-1}. Morphology was analyzed by field emission scanning electron microscopy (FE-SEM, NOVA NanoSEM 450, American FEI Company, Hillsboro, OR, USA). Transmission electron microscopy (TEM) and high-resolution transmission electron microscopy (HRTEM) analysis were performed (Tecnai G2 F30S, American FEI Company). X-ray photoelectron spectroscopy (XPS) analysis was performed using ESCALAB 250 X-ray photoelectron spectroscopy (Thermo Fisher Company, Waltham, MA, USA) with Al Kα radiation.

2.4. Electrocatalytic Measurements

A three-electrode system was used for electrocatalytic test at room temperature (298.15 K). A Pt sheet was used as the counter electrode, the electrode Ag/AgCl was used as a counter reference. Ni_xCo_y-LDH@Cs were used as the working electrodes, relative to the standard hydrogen electrode (RHE). Voltage: $E_{RHE} = E_{Ag/AgCl} + 0.197\ V + 0.059 \times pH$. Linear sweep voltammetry (LSV) was used for the polarization curve of the electrochemical catalyst. The scanning rate was 2 mV/s. Stable tests were performed at the current density of 10 mA cm^{-2}. Electrochemical impedance spectroscopy (EIS) was used to measure the frequency range 0.1~100 kHz. Cyclic voltammetry (CV, scanning rate 50 mV/s) and the timing current method were used for the stability test. Electrochemical double-layer capacitance (C_{dl}) was measured by cyclic voltammetry curves at different scanning rates: CV test voltage range −0.05~0.45V (vs. RHE); scanning rate 5~100 mV/s. C_{dl} was calculated according to the following formula: ic = VC_{dl}. Here, ic is the exchange current (mA/cm^2), V is the scanning rate (mV/s), and C_{dl} is the double layer capacitance (F/cm^2). The turnover frequency range (TOF) was calculated at potentials of 1.0~1.5 V. The supporting information provides elaborate descriptions of the test methods employed for infrared spectroscopy, the techniques utilized for determining the total solution of water, and the calculation methods applied for density functional theory (DFT).

3. Results

3.1. Characterization of Electrocatalysts

Ni_xCo_y-LDH@C was prepared using the condensate reflux method. Then, NiCo-LDH@C was prepared by injecting nitrogen and calcining it at 800 °C. By adjusting the ratio of Ni/Co, catalysts with different Ni contents can be prepared (Figure 1a). They are labeled Ni_2Co-LDH@C, $NiCo_2$-LDH@C, and NiCo-LDH@C, according to the ratio of Ni/Co (Figure 1b). It can be seen that the main diffraction peaks of the sample before and after etching basically match the standard XRD cards (JCPDS:46-0943, 26-0408). The diffraction peaks of samples at 11.1°, 35.4°, and 48.9° are different from those of (003), (012), and (110), respectively. After loading with silk fibroin, diffraction peaks corresponding to the (003), (012), and (110) crystal planes of the sample still exist, and there are obvious diffraction peaks corresponding to silk fibroin at 31.2° and 52.5°, indicating that Ni_2Co-LDH was successfully loaded onto the silk and increased the layer spacing of LDH [23,24]. To understand the chemical valence states of each element in Ni_2Co−LDH@C, the XPS of Ni_2Co-LDH@C is given in Figure 2. The survey spectrums of Ni_2Co-LDH@C showed that the sample material contains only Ni, Co, O, and C elements (Figure 1c). The XPS quantitative analysis results, as presented in Table S1, reveal Ni:Co ratios of 2.14:1.54 in NiCo-LDH@C, 5.43:2.45 in Ni_2Co-LDH@C, and 2.9:4.86 in $NiCo_2$-LDH@C, which impeccably align with the molar ratio indicated by the molecular formula. In the Ni_{2p} emission spectrum (Figure 1d), in addition to the two obvious satellite peaks, there are diffraction peaks with binding energies of 855.5 eV and 873.3 eV, corresponding to $Ni_{2p3/2}$ and $Ni_{2p1/2}$, respectively, indicating the existence of only Ni^{2+}. Gaussian fitting Co_{2p} high-resolution spectrum includes two satellite peaks and two spin orbit peaks (Figure 1e). In Co_{2p}, the peak values of binding energy at 780.5 eV and 796.2 eV correspond to Co^{3+}, and the binding energies at 782.3 eV and 796.6 eV belong to Co^{2+}. The appearance of Co^{3+} may be caused by the effect of oxygen in solution on the sample at high temperature [23–26]. The distribution of C in XPS data was roughly as follows: C-C 284.9 eV, C=O 288.3 eV, C-O 287.5 eV (as shown in Figure S1). IR spectra of Ni_2Co-LDH@C are shown in Figure 1f. The spectrum contained absorption vibration peaks of three types of functional groups. Two absorption peaks of 3426 and 1630 cm^{-1} were caused by stretching vibration of hydroxyl O-H on the laminate and bending vibration of interlaminar water molecules. The strong absorption peak of 1404 cm^{-1} can be attributed to the symmetric tensile vibration of Co-O, which indicates that the Co-O intercalationthe low absorption peak is caused by lattice vibration of metal–oxygen bond (Ni-O or Co-O) [27–30]. Figure 1g shows the Raman spectra of Ni_2Co-LDH@C. The characteristic peaks of 1342, 1575, and 884 cm^{-1} in the Raman spectrum correspond to D, G, and 2D peaks of the silk carbonized material, respectively. Among them, the D peak is the defect peak, which was caused by structural defects and disorder. In addition, the Raman spectra of the grown Ni_2Co-LDH@C materials demonstrate the vibration peaks of the Raman displacement at 459 and 517 cm^{-1}, which correspond to the tensile vibration of the Ni-OH bond and the symmetric stretching vibration mode of A_g in the Co-O bond, respectively [31,32]. In addition, by observing the Raman characteristic peak of the composite sample, the Ni_2Co-LDH@C composite structure was successfully synthesized.

In order to play a more efficient role, the conductivity of each carbonized silk fiber must be significantly enhanced. The current received by each component is required to efficiently conduct and bundle through the substrate material, thereby increasing its efficiency. Nickel and cobalt were loaded onto the silk by condensation reflux and then carbonized and modified with Ni_2Co-LDH. Efficient electrostatic materials were fabricated with better electrical conductivity, richer surface structure, and higher stability and flexibility than conventional substrate materials. Self-supporting structural systems give them richer means of modification [33,34]. Ni_2Co-LDH@C showed the SEM of two-dimensional rod-like morphology (Figure 2a,b), and aggregated Ni_2Co-LDH particles were observed on the surface of carbonized silk (Figure 2c). In addition, the atomic structure of Ni_2Co-LDH was revealed by HRTEM. The Ni and Co sites of the coordination were found to be uniformly anchored throughout the surface (Figure 2c). HRTEM was used to the study structural changes in the loaded carbonized silk. Figure 2e showed shows that the crystal spacing of Ni_2Co-LDH is 0.242 nm and 0.353 nm for the (111) and (002) crystal planes of Ni_2Co-LDH, respectively. Figure 2f shows that the two groups of nicely resolved lattice fringes at 0.353 nm and 0.261 nm belong to the (002) plane of NiCo-LDH, and 0.261 nm corresponds to the (012) crystal plane. The Ni_2Co-LDH@C with carbon atoms preserves the crystal structure of the NiCo-LDH, in agreement with the XRD results [35]. The EDS indicated a continuous decrease in the Ni content with the introduction of more Co atoms, indicating that some Ni atoms are replaced by Co atoms. The Wt% values of the Ni_xCo_y-LDH@C EDS analysis are presented in Table S2, showing the compositional ratios of NiCo-LDH@C (Ni/Co = 27.16:28.56), Ni_2Co-LDH@C (Ni/Co = 40.98:18.37), and $NiCo_2$-LDH@C (Ni:Co = 15.98:33.17). The Wt% analysis revealed that the molar ratios of Ni to Co are 1:1, 2:1, and 1:2, respectively, which closely align with the XPS result. In electrochemical performance tests, Ni_2Co-LDH@C was found to have the best catalytic performance (Figure 2g–i).

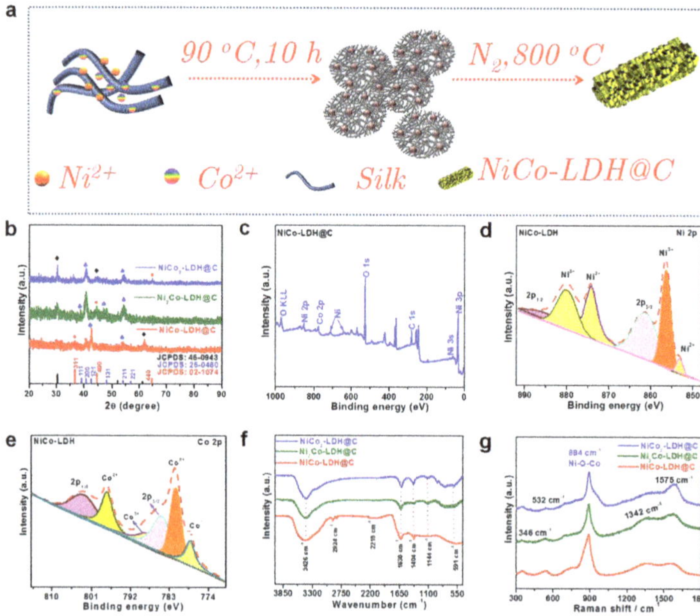

Figure 1. (**a**) Schematic diagram of the Ni_2Co-LDH@C preparation via a two-step method. (**b**) XRD pattern of the Ni_xCo_y-LDH@C; (**c**) XPS survey of the Ni_2Co-LDH@C catalysts; (**d**) the spectra of Ni_{2p}; (**e**) the spectra of Co_{2p}; (**f**) the FT-IR spectra of the Ni_xCo_y-LDH@C catalysts; (**g**) Raman spectra of the Ni_xCo_y-LDH@C catalysts.

3.2. Electrocatalytic Performance for the HER

The HER of the catalysts was studied in 1 mol/L KOH solution. Figure 3a showed NiCo-LDH@C, Ni_2Co-LDH@C, $NiCo_2$-LDH@C, and Pt/C. As expected, Ni_2Co-LDH@C showed the highest HER activity among the three synthesized catalysts, and the current density of 10 mA cm^{-2} only needed 164.8 mV overpotential. However, $NiCo_2$-LDH@C and NiCo-LDH@C require 232 and 181.6 mV, respectively. Figure 3b shows the corresponding EIS for these catalysts. The R_{ct} of Ni_2Co-LDH@C was 3.32 Ω, indicating that the interface electron transfer rate of Ni_2Co-LDH@C was faster, and it could maintain a low overpotential even at a high current density. The Tafel of Ni_2Co-LDH@C was 113.9 mV dec^{-1}, which was lower than $NiCo_2$-LDH@C (242.8 mV dec^{-1}) and NiCo-LDH@C (230.49 mV dec^{-1}), 20 wt% Pt/C (38.7 mV dec^{-1}) (Figure 3c). Under the same current density of 10 mA cm^{-2}, the Ni_2Co-LDH@C catalyst exhibited significantly higher current density compared to catalysts with different molar ratios of components and outperformed commercial 20 wt% Pt/C catalysts. These results indicate that the prepared catalyst with a molar ratio of 2:1 demonstrates superior electrocatalytic activity for HER. The results showed that $NiCo_2$-LDH@C had the fastest reaction rate. In addition, Ni_2Co-LDH@C performance was better than many Ni-based catalysts (Table S3). These results indicated that the interfacial effect of Ni/Co heterostructure and the three-dimensional structure composed of carbon materials were conducive to enhancing the hydrogen evolution activity of the catalysts [36,37]. In order to compare the intrinsic activity of the catalysts, the electrochemical active surface area (ECSA) of catalytic activity was normalized. ECSA was proportional to the double layer capacitance (C_{dl}) of the catalyst (Figure S2), using CV curves at different scanning rates, the C_{dl} values of the NiCo-LDH@C, $NiCo_2$-LDH@C, and Ni_2Co-LDH@C can be calculated as 4.65 mF cm^{-2}, 3.80 mF cm^{-2}, and 26.10 mF cm^{-2}, respectively. Ni_2Co-LDH@C had a lower activation energy of only 6.6 kJ/mol compared to the remaining two catalysts. It was considerably smaller than $NiCo_2$-LDH@C (11 kJ/mol), NiCo-LDH@C (7.2

kJ/mol), and Ni_2Co-LDH (12.8 kJ/mol), and the adsorption of hydrogen by Co is overly strong, resulting the low activity of electrocatalytic decomposition of aquatic hydrogen. Comparing the electrocatalytic activities of several catalysts with current densities of 10 mA·cm^{-2}, 50 mA/cm^{-2}, and 100 mA·cm^{-2}, Ni_2Co-LDH@C had the lowest overpotential (Figure 3f). The Ni_2Co-LDH@C catalyst had the lowest hydrogen evolutionary excess potential compared to similar published catalysts. To verify the stability of the sample, after circulating the catalyst for 1000 cycles, the coincidence degree of the curve elevated and practically no shift is present, indicating perfect stability. As shown in Figure 3i, at the initial current density of 10 mA·cm^{-2}, the sample showed no obvious attenuation after 12 h.

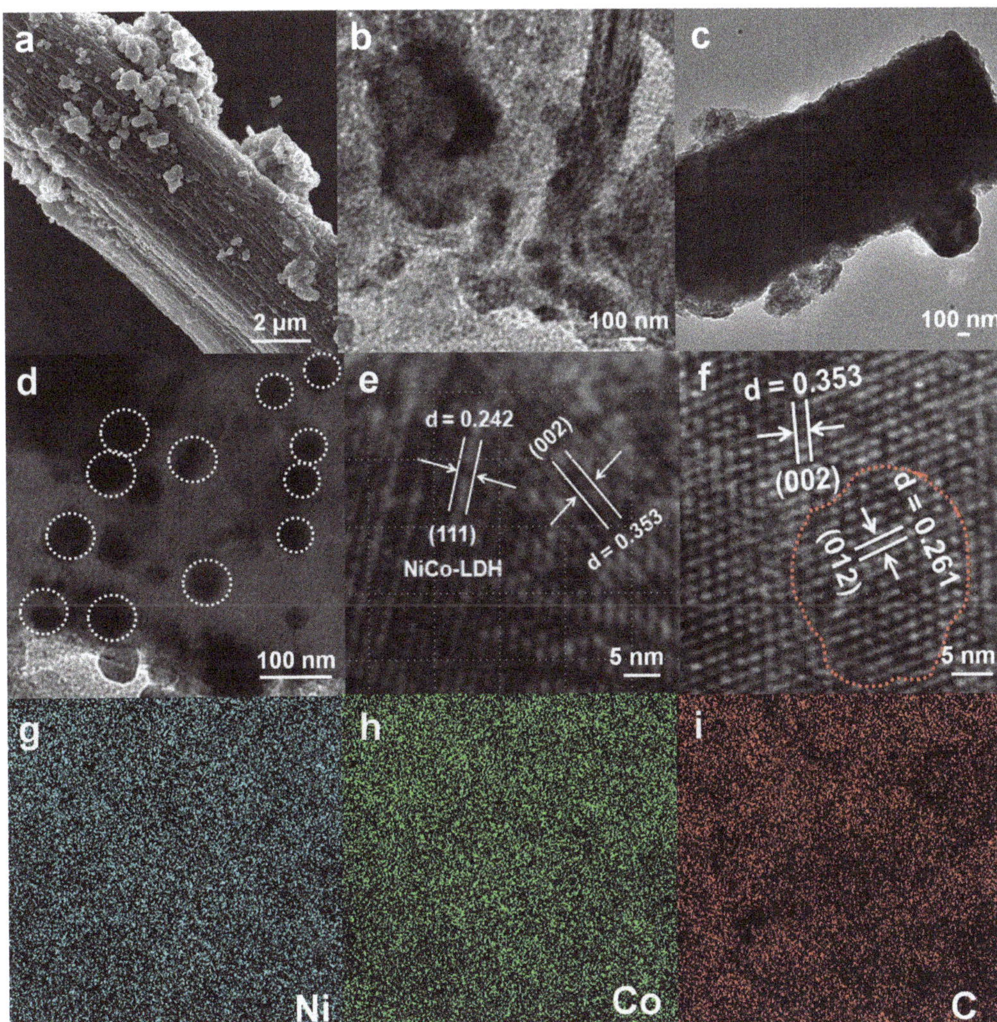

Figure 2. (**a**) SEM of the Ni_2Co-LDH@C catalysts. Scale bar: 2 μm. (**b**) TEM of the carbonization of silk. (**c**,**d**) TEM of the Ni_2Co-LDH@C catalysts. Scale bar: 100 nm. (**e**) HRTEM of the Ni_2Co-LDH catalysts. (**f**) HRTEM of the Ni_2Co-LDH@C catalysts. Scale bar: 5 nm. (**g**–**i**) Mapping of Ni, Co, and C elements.

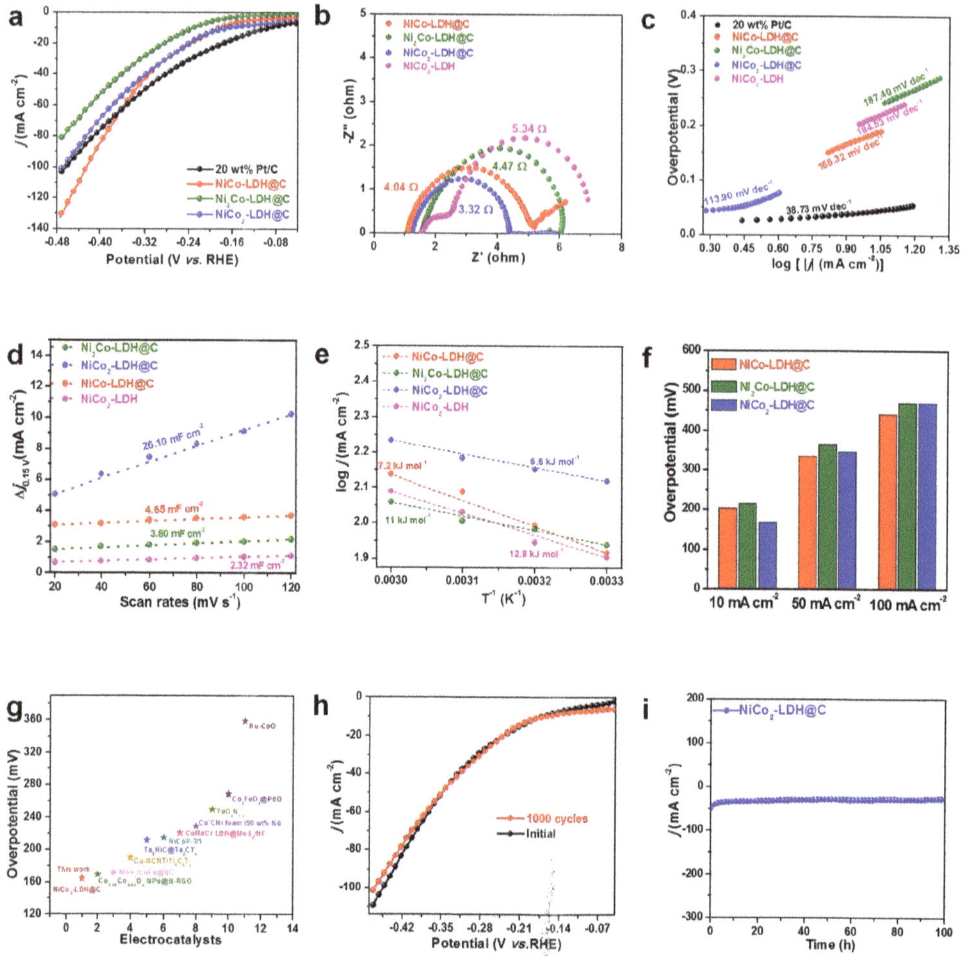

Figure 3. (**a**) HER LSV curves of NiCo-LDH@C, Ni_2Co-LDH@C, and $NiCo_2$-LDH@C at 20 wt% Pt/C/NF. (**b**) EIS tests of NiCo-LDH@C, Ni_2Co-LDH@C, and $NiCo_2$-LDH@C. (**c**) Tafel plots of NiCo-LDH@C, Ni_2Co-LDH@C, and $NiCo_2$-LDH@C. (**d**) ECSA of NiCo-LDH@C, Ni_2Co-LDH@C, and $NiCo_2$-LDH@C. (**e**) Ea of NiCo-LDH@C, Ni_2Co-LDH@C, and $NiCo_2$-LDH@C. (**f**) The comparisons of overpotential at 10 mA cm^{-2} (η_{10}) and 50 mA cm^{-2} of NiCo-LDH@C, Ni_2Co-LDHs@C, and $NiCo_2$-LDHs@C. (**g**) HER performance comparison. (**h**) Polarization curves for the Ni_2Co-LDH@C before and after 1000 cycles. (**i**) The chronoamperometric curves of Ni_2Co-LDH@C for 100 h.

3.3. Electrocatalytic Performance for the OER

The OER catalytic activity of the samples was evaluated in the 1.0 mol/L KOH electrolyte. Figure 4a shows the polarization curves of samples obtained by linear sweep voltammetry (LSV). As can be seen from the polarization curves, the OER catalytic performance of the samples obtained is gradually enhanced with the difference of Ni/Co ratio. When Ni:Co = 2:1, it shows the optimal catalytic activity. When the current density is 10 mA·cm^{-2}, the overcurrent potential is only 183.4 mV. Ni_2Co-LDH@C performance was better than many Ni-based catalysts (Table S4). However, when Ni:Co = 1:1, the catalytic activity of the obtained sample suddenly decreased, which may be due to the collapse of the NiCo-LDH structure, resulting in a sharp reduction in the specific surface area. The interfacial charge transfer resistance (R_{ct}) controls the charge transfer efficiency of interfacial reaction. As shown in Figure 4b, the R_{ct} of Ni_2Co-LDH@C was 2.95 Ω, indicating that the interface electron transfer rate of Ni_2 Co-LDH@C was faster in the OER reaction. The Tafel slope reflected the kinetic performance of OER. In general, the smaller the Tafel slope is, the better the kinetic performance of the catalyst. As can be seen from Figure 4c, the slope of Tafel decreased from 169 mV dec^{-1} to 108.27 mV dec^{-1} with a different Ni:Co ratio, indicating that the introduction of Ni ions can improve the dynamic performance of Ni_2Co-LDH@C. In order to compare the intrinsic activity of the catalysts, the Ea of catalytic activity was normalized, as shown in Figure 4d. The defect-rich structure and metal/oxide interface substantially improved the intrinsic activity of the catalyst: its Ea was Ni_2Co-LDH@C < $NiCo_2$-LDH@C < NiCo-LDH@C < Ni_2Co-LDH. Compared with similar published catalysts, the Ni_2Co-LDH@C catalyst had the lowest oxygen evolution overpotential (Figure 4e). The OER overpotential was compared at different current densities (10, 50, 100 mA·cm^{-2}). The results showed that the Ni_2Co-LDH@C catalyst had the lowest excess potential at different current densities, suggesting that Ni_2Co-LDH@C had better catalytic activity for oxygen evolution (Figure 4f). Figure 4g shows the multistep timing voltage analysis curves; similarly to the results of OER, its voltage does not change significantly as the current density increases sequentially from −30 to −300 mA·cm^{-2}, and each time it increases by −30 mA cm^{-2} and stays for 500 s. By coupling Ni_2Co-LDH@C, $NiCo_2$-LDH@C, and NiCo-LDH@C for the heterogeneous structure, Ni:Co not only makes use of the bifunctional properties, but also improves the intrinsic activity of the catalyst through the interface effect [38,39]. In addition to catalytic activity and kinetic properties, stability is also an important factor in evaluating the performance of catalysts. The stability of Ni_2Co-LDH@C characterized by loop test, after 1000 cycles, the curves basically coincide (Figure 4h). At the initial current density of 10 mA·cm^{-2}, the current density of Ni_2Co-LDH@C can remain after 12 h (Figure 4i), indicating that it has excellent long-term stability in OER catalytic reaction.

3.4. Overall Water Splitting Performance

Ni_2Co-LDH@C||Ni_2Co-LDH@C were used as both cathode and anode catalysts for the water splitting reaction. Figure 5a showed the polarization curves of its catalyst reaction. Ni_2Co-LDH@C||Ni_2Co-LDH@C required only 1.63, 1.82, and 1.94 V to obtain current densities of 10, 50, and 100 mA cm^{-2}, respectively (Figure 5b). In addition, the performance of the Ni_2Co-LDH@C||Ni_2Co-LDH@C catalyst was better than others reported (Figure 5c). The above results indicate that Ni_2Co-LDH@C||Ni_2Co-LDH@C had excellent HER and OER stability. The intrinsic catalytic performance of Ni_2Co-LDH@C||Ni_2Co-LDH@C electrode was evaluated by oxygen turnover frequency. The cyclic voltammetry diagram showed the linear relationship with the scanning rate. The cyclic voltammetry (Figures S3–S5) showed that the oxidation peak of the sample had a linear relationship with the scanning rate, after fitting with LSV (Figure 5d). Table S5 shows that the oxygen turnover frequency of NiCo-LDH@C was 0.0103 s^{-1} and that of $NiCo_2$-LDH@C was 0.0216 s^{-1}. Ni_2Co-LDH@C reached 0.0597 mol O_2 s^{-1} under the overpotential of 190.6 mV. Table S6 shows the oxygen turnover frequency of $NiCo_2$-LDH@C||$NiCo_2$-LDH@C, which reached 0.0615 mol O_2 s^{-1} under the overpotential of 181.6 mV (Figure 5e). The catalysts with a three-dimensional array structure supported on the loose and porous surface of the carbon materials facilitated the exposure of the active site and the escape of bubbles as well as contact with the electrolyte. Under the same conditions, the performance of nickel-based transition metal electrocatalysts is better than most of the previously reported nickel-based transition metal electrocatalysts (Table S7). The voltage test experiment (Figure 5h) shows that the Ni_2Co-LDH@C||Ni_2Co-LDH@C electrodes had good stability of total solution water at 1.66 V, and the potential change can be ignored. Figure 5f shows the overall water splitting performance of the device over 100 h (producing O_2) and the cathode (producing H_2); the two-electrode catalyst showed good stability.

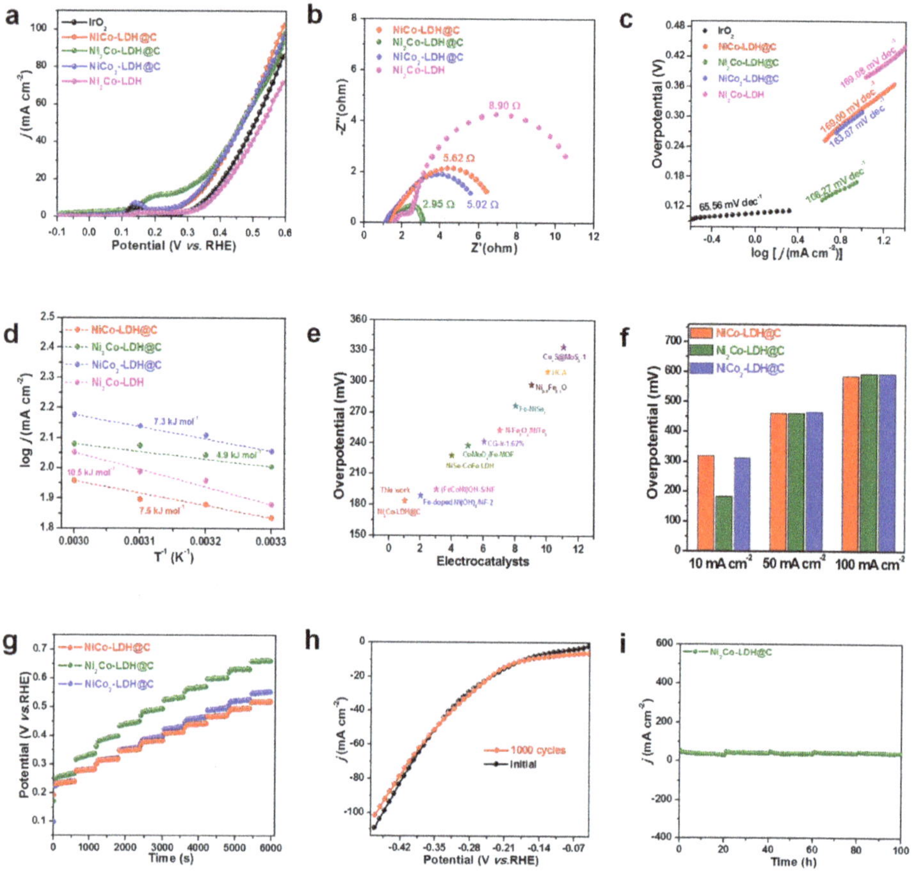

Figure 4. (a) OER LSV curves of NiCo-LDH@C, Ni_2Co-LDH@C, $NiCo_2$-LDH@C, and IrO_2. (b) EIS curves of NiCo-LDH@C, Ni_2Co-LDH@C, $NiCo_2$-LDH@C, and IrO_2. (c) Tafel plots of NiCo-LDH@C, Ni_2Co-LDH@C, and $NiCo_2$-LDH@C. (d) Ea of NiCo-LDH@C, Ni_2Co-LDH@C, and $NiCo_2$-LDH@C. (e) OER performance comparison. (f) The comparisons of overpotential at 10 mA cm^{-2} (η_{10}) and 50 mA cm^{-2} of NiCo-LDH@C, Ni_2Co-LDH@C, and $NiCo_2$-LDH@C. (g) CP plots on the catalyst. (h) Polarization curves for the Ni_2Co-LDH@C before and after 1000 cycles. (i) The chronoamperometric curves of Ni_2Co-LDH@C for 100 h.

3.5. Theoretical Calculation Analysis

Ni_2Co-LDH@C had a higher electron density than Ni_2Co-LDH in the Fermi level, indicating that Ni_2Co-LDH@C had faster electron transfer, thereby improving its electrical conductivity and showing enhanced catalytic activity, which was consistent with experimental tests (Figure 6a). The C is introduced into Ni_2Co-LDH in alkaline media to improve HER performance. This was investigated using DFT models. The adsorption of H_2O on catalyst surface and the cleavage of H-OH bond with OH* and H* were the key steps of the HER reaction under alkaline conditions. The mechanism of the excellent HER properties was investigated by using the Gibbs free energies of H_2O(ΔG_{H2O}) and H*(ΔG_{H^*}). The ΔG_{H^*} value of Ni_2Co-LDH@C was tiny, indicating that the supporting carbon materials were conducive to improving HER performance (Figure 6b). The responsible for the excellent HER performance was explained in Figure 6c. Since the weak adsorption of HER to intermediates is the key factor for its elevated activity, the physical mechanism of hydrogen evolution discussed by studying the electronic structure of Ni_2Co-LDH@C (Figure 6c). In general, the

OER pathway is a four-electron transfer step, including: $H_2O \rightarrow *OH + (H^+ + e^-)$, $*OH \rightarrow *O + (H^+ + e^-)$, $*O + H_2O^- \rightarrow *OOH + (H^+ + e^-)$ and $*OOH \rightarrow O_2 + (H^+ + e^-)$ (Figure 6d) [40].

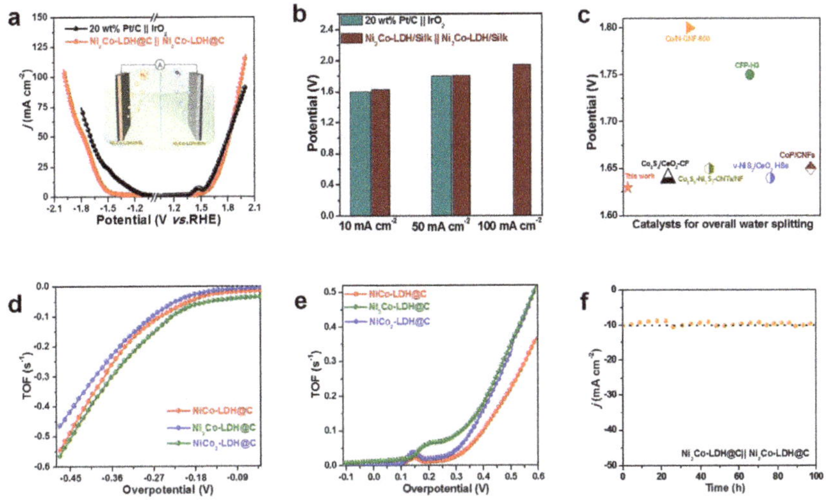

Figure 5. (**a**) LSV curves of 20 wt% Pt/C | | IrO_2 and Ni_2Co-LDH@C | | Ni_2Co-LDH@C in a two-electrode configuration in 1 M KOH. (**b**) The comparisons of potential at 10 mA cm^{-2}, 50 mA cm^{-2} and 100 mA cm^{-2} of Ni_2Co-LD@C | | Ni_2Co-LDH@C. (**c**) Comparison performance of overall water splitting. (**d**) The calculated TOF values of the NiCo-LDH@C, Ni_2Co-LDH@C, and $NiCo_2$-LDH@C varied with the HER potentials. (**e**) The calculated TOF values of the NiCo-LDH@C, Ni_2Co-LDH@C, and $NiCo_2$-LDH@C varied with the OER potentials. (**f**) The stability measurements of Ni_2Co-LDH@C | | Ni_2Co-LDH@C in 1 M KOH.

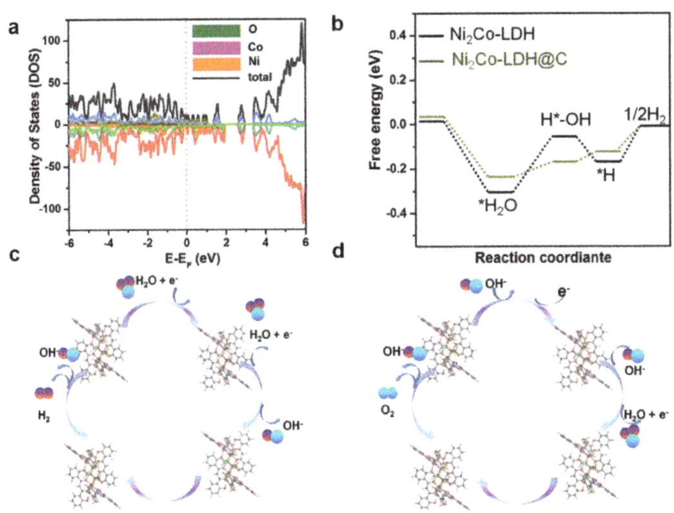

Figure 6. (**a**) DOS of Ni_2Co-LDH@C; (**b**) Gibbs free-energy diagrams of HER intermediates for Ni_2Co-LDH@C; (**c**) Schematic diagram of HER mechanism; (**d**) Schematic diagram of OER mechanism.

4. Conclusions

In summary, a novel electrocatalytic synthesis strategy has been described for the self-sustained stabilization of defective NiCo-LDH-active sites in carbon materials. Based on the precise tuning of the active site Ni:Co, Ni_2Co-LDH@C offers superior HER and OER properties with low excess potential, elevated current density, and long-term persistence in alkaline media. The well-defined structure of the catalyst also allows for a fundamental study of the reaction steps and dynamics of the HER and OER reactions. For example, DFT calculations show that Ni_2Co-LDH@C promotes favorable regulation of the H_2 adsorption energy by HER and O-O coupling to the OER. This work not only develops a simple and practical strategy for the synthesis of Ni_2Co-LDH@C, but also theoretically/experimentally confirms the key role of carbon materials as self-supporting catalysts in the unexpected optimization of electrocatalytic activity, opening up new opportunities for the development of efficient and stable electrocatalysts, and potentially for the development of improved commercial water splitting processes.

Supplementary Materials: The following supporting information can be downloaded at: https://www.mdpi.com/article/10.3390/c9040094/s1, Figure S1: XPS spectra of C_{1s} XPS spectra; Figure S2: Cyclic voltammograms of Ni_xCo_y-LDH@C; Figures S3–S5: Cyclic voltammograms of Ni_xCo_y-LDH@C; Table S1: Comparison of HER performance; Table S2: Comparison of OER performance; Table S3: OER-TOF of the catalyst at the specific overpotential; Table S4: OER-TOF of the catalyst at the specific overpotential; Table S5: Comparison of overall spiltting water performance. Table S6: HER-TOF determined at 167.6 mV (vs. RHE). Table S7: The overall performances of the as-prepared electrode and other electrodes with electrocatalysts in 1.0 M KOH.

Author Contributions: Conceptualization, Z.W. also processed the experimental data, performed the analysis. P.T. partially contributed to the collection of data. L.D. processed the experimental data. Y.Y. processed the experimental data. L.L. and J.L. funding acquisition. M.Y. was involved in the planning, formulation, design, and implementation of the research. G.D. formulation, design, and implementation of the research. All authors have read and agreed to the published version of the manuscript.

Funding: This research received the Open Foundation of Key Laboratory of Auxiliary Chemistry and Technology for Chemical Industry, Ministry of Education, Shaanxi University of Science and Technology (KFKT2021-06), National Undergraduate Training Program for Innovation and Entrepreneurship (S202314389160). Supported by the Opening Project of Structural Optimization and Application of Functional Molecules Key Laboratory of Sichuan Province (2023GNFZ-02). Medical Imaging Key Laboratory of Sichuan Province, North Sichuan Medical College, Nanchong, Sichuan, China, 637000 (MIKL202204).

Data Availability Statement: Data are contained within the article.

Acknowledgments: D/MAX2400 X-ray diffrotometer produced by Company in Japan was adopted for X-ray diffraction (XRD) analysis, with CuKa (X = 1.54 Å) is the radiation source; operating voltage was 200 kV, current was 50 mA, 2θ is 10–80°, and the scanning rate was 10 min^{-1}. Raman spectroscopy (Thermo Fisher Company in the United States); laser wavelength was 532 nm and the detection range was 0–2500 cm^{-1}. Morphology was analyzed by field emission scanning electron microscopy (FE-SEM, NOVA NanoSEM 450, American FEI Company). Transmission electron microscope (TEM) and high-resolution transmission electron microscope (HRTEM) analysis used by Tecnai G2 F30S (American FEI company). X-ray photoelectron spectroscopy (XPS) analysis was performed using ESCALAB 250 X-ray photoelectron spectroscopy (Thermo Fisher Company) with Al Kα radiation.

Conflicts of Interest: The authors declare no conflict of interest.

References

1. Sun, T.; Tang, Z.Y.; Zang, W.J.; Li, Z.J.; Li, J.; Li, Z.H.; Cao, L.; Rodriguez, J.S.D.; Mariano, C.O.M.; Xu, H.; et al. Ferromagnetic single-atom spin catalyst for boosting water splitting. *Nat. Nanotechnol.* **2023**, *18*, 763–771. [CrossRef] [PubMed]
2. Zang, Y.; Lu, D.Q.; Wang, K.; Li, B.; Peng, P.; Lan, Y.Q.; Zang, S.Q. A pyrolysis-free Ni/Fe bimetallic electrocatalyst for overall water splitting. *Nat. Commun.* **2023**, *14*, 1792. [CrossRef] [PubMed]
3. Zhu, J.; Du, Q.X.; Khan, M.A.; Zhao, H.B.; Fang, J.H.; Ye, D.X.; Zhang, J.J. 2D porous Co-Mo nitride heterostructures nanosheets for highly effective electrochemical water splitting. *Appl. Surf. Sci.* **2023**, *623*, 156989. [CrossRef]
4. Kharabe, G.P.; Illathvalappil, R.; Barik, S.; Kanheerampockil, F.; Walko, P.S.; Bhat, S.K.; Devi, R.N.; Kurungot, S. A cobalt–manganese modified theophrastite phase of nickel hydroxide nanoflower arrays on nickel foam as a self-standing bifunctional electrode for overall water electrolysis. *Sustain. Energy Fuels* **2023**, *7*, 2428–2440. [CrossRef]

5. Liu, Z.Y.; Li, Y.; Peng, S.S.; Huang, J.; Su, B.L.; Gao, K.; Zhou, S.; Gao, Z.J.; Chen, T. A comprehensive melamine sponge decorated with 3D flowerlike Ag/NiCo-layered double hydroxide for wastewater remediation. *J. Environ. Chem. Eng.* **2022**, *10*, 108447. [CrossRef]
6. Wijaya, R.A.; Suseno, A.; Lusiana, R.A.; Septina, W.; Harada, T. Synthesis of $CuInS_2$ thin film photocathode with variation of sulfurization sources and Pt-In_2S_3 modification for photoelectrochemical water splitting. *J. Electroanal. Chem.* **2023**, *945*, 117683.
7. Hu, Y.H.; Shen, T.Y.; Song, Z.H.; Wu, Z.H.; Bai, S.; Liu, G.H.; Sun, X.L.; Wang, Y.Y.; Hu, S.Y.; Zheng, L.R.; et al. Atomic Modulation of Single Dispersed Ir Species on Self-Supported NiFe Layered Double Hydroxides for Efficient Electrocatalytic Overall Water Splitting. *ACS Catal.* **2023**, *13*, 11195–11203. [CrossRef]
8. Lee, D.E.; Moru, S.; Reddy, K.P.; Jo, W.K.; Tonda, S. Bimetallic Cu–Ni core–shell nanoparticles anchored N-doped reduced graphene oxide as a high-performance bifunctional electrocatalyst for alkaline water splitting. *Appl. Surf. Sci.* **2023**, *622*, 156928. [CrossRef]
9. Liu, Z.; Liu, Y.X.; He, H.W.; Shao, H.B.; Zhang, Y.; Li, J.; Cai, W.W. Valence regulation of Ru/Mo_2C heterojunction for efficient acidic overall water splitting. *Electrochim. Acta* **2023**, *443*, 141920. [CrossRef]
10. Nguyen, T.H.T.; Zarkua, Z.; Chinnappa, C.V.; Hu, W.J.; Sree, S.P.; Grandjean, D.; Pant, D.; Janssens, E. $Co_{3-x}Fe_xO_4$ inverse opals with tunable catalytic activity for high-performance overall water splitting. *Nanoscale* **2023**, *15*, 10306–10318. [CrossRef]
11. Luo, J.B.; Zhou, Y.; Wang, X.Z.; Gu, Y.F.; Liu, W.L.; Wang, S.T.; Zhang, J. $CoMoO_4$-CoP/NC heterostructure anchored on hollow polyhedral N-doped carbon skeleton for efficient water splitting. *J. Colloid Interface Sci.* **2023**, *648*, 90–101. [CrossRef]
12. Pan, Y.D.; Gao, J.K.; Lv, E.J.; Li, T.T.; Xu, H.; Sun, L.; Nairan, A.; Zhang, Q.C. Integration of Alloy Segregation and Surface Co–O Hybridization in Carbon-Encapsulated CoNiPt Alloy Catalyst for Superior Alkaline Hydrogen Evolution. *Adv. Funct. Mater.* **2023**, *33*, 2303833. [CrossRef]
13. Jing, F.L.; Pi, Y.Y.; Zhao, C.X.; Zhou, H.; Luo, S.Z.; Fang, W.H. Constructing highly stable dendritic silica-wrapped NiCo/carbon nanotubes architecture for hydrogen production. *Mater. Today Sustain.* **2022**, *19*, 100183. [CrossRef]
14. Liu, B.H.; Li, S.Y.; Wang, T.Q.; Yang, Y.; Wang, L.X.; Zhang, X.; Liu, Z.C.; Niu, L.W. Construction of CoFe bimetallic phosphide microflowers electrocatalyst for highly efficient overall water splitting. *Catal. Commun.* **2023**, *175*, 106607. [CrossRef]
15. Li, N.; Qu, S.J.; Qin, F.F.; Li, H.Y.; Shen, W.Z. Construction of P,N-codoped carbon shell coated CoP nanoneedle array with enhanced OER performance for overall water splitting. *Int. J. Hydrogen Energy* **2023**, *48*, 27215–27230. [CrossRef]
16. Li, N.; Qu, S.J.; Ma, J.J.; Shen, W.Z. Core-shell-structured CoS_2@N-doped carbon nanoneedle array as an efficient bifunctional electrocatalyst for overall water splitting. *Int. J. Hydrogen Energy* **2023**, *48*, 180–195. [CrossRef]
17. Yang, M.; Cui, Z.; Wang, Q.H.; Yin, J.F.; Yang, L.; Wang, O.; Deng, G.W. Improvement of specific capacitance and rate performance of $NiWO_4$ synthesized through modified chemical precipitation. *J. Mater. Sci. Mater. Electron.* **2021**, *32*, 12232–12240. [CrossRef]
18. He, X.D.; Han, X.; Zhou, X.Y.; Chen, J.D.; Wang, J.; Chen, Y.; Yu, L.H.; Zhang, N.; Li, J.; Wang, S.; et al. Electronic modulation with Pt-incorporated NiFe layered double hydroxide for ultrastable overall water splitting at 1000 mA cm^{-2}. *Appl. Catal. B Environ.* **2023**, *331*, 122683. [CrossRef]
19. Chen, K.; Cao, Y.H.; Yadav, S.; Kim, G.C.; Han, Z.; Wang, W.M.; Zhang, W.J.; Dao, V.; Lee, I. Electronic structure reconfiguration of nickel–cobalt layered double hydroxide nanoflakes via engineered heteroatom and oxygen-vacancies defect for efficient electrochemical water splitting. *Chem. Eng. J.* **2023**, *463*, 142396. [CrossRef]
20. Peng, M.W.; Zhao, B.; Shi, D.L.; Wang, Y.W.; Li, D.; Liang, W.K.; Yang, H.; Liang, Z.Q.; Sun, Y.H.; Jiang, L. Enabling high loading of well-dispersed Ni_2CoP_2 catalysts on a 3D-printed electrode for efficient electrocatalysis. *J. Mater. Chem. A* **2023**, *11*, 15394. [CrossRef]
21. Li, G.L.; Miao, Y.Y.; Qiao, X.Y.; Wang, T.Y.; Deng, F. Engineering edge sites based on NiS_2/MoS_2/CNTs heterojunction catalyst for overall water splitting. *Appl. Surf. Sci.* **2023**, *615*, 156309. [CrossRef]
22. Fei, L.S.; Sun, H.N.; Xu, X.M.; Li, Y.; Ran, R.; Zhou, W.; Shao, Z.P. Understanding the bifunctional catalytic ability of electrocatalysts for oxygen evolution reaction and urea oxidation Reaction: Recent advances and perspectives. *Chem. Eng. J.* **2023**, *471*, 144660. [CrossRef]
23. Yang, H.S.; Fu, Q.; Hou, Y.; Yu, J.X.; Tong, Y.J.; Li, N.; Zhong, H.J.; Zhu, F.; Wang, J.H.; Hao, Z.P.; et al. Ultrahigh surface area of single-atom iron nanosheets assists in the efficient utilization of reactive oxygen species in the peroxymonosulfate activation process for pollutant removal. *Environ. Sci. Nano* **2023**, *10*, 1152. [CrossRef]
24. Huang, Q.P.; Jiang, M.J.; Li, Y.Y.; Liang, C.; Tang, Y.M.; Xie, F.Y.; Yang, M.; Deng, G.W. Construction of $MnxCoyO_4$/Ti electrocatalysts for efficient bifunctional water splitting. *Dalton Trans.* **2022**, *51*, 9085. [CrossRef]
25. Sun, A.; Qiu, Y.L.; Wang, Z.X.; Cui, L.; Xu, H.Z.; Zheng, X.Z.; Xu, J.T.; Liu, J.Q. Interface engineering on super-hydrophilic amorphous/crystalline NiFe-based hydroxide/selenide heterostructure nanoflowers for accelerated industrial overall water splitting at high current density. *J. Colloid Interface Sci.* **2023**, *650*, 573–581. [CrossRef]
26. Qiu, S.P.; Zhang, B.J.; Wang, X.; Huang, J.Z.; Zhao, G.; Ding, M.; Xu, X.J. Interface strong-coupled 3D Mo-NiS@Ni-Fe LDH flower-cluster as exceptionally efficient electrocatalyst for water splitting in wide pH range. *J. Colloid Interface Sci.* **2023**, *641*, 277–288. [CrossRef]
27. Nayem, S.M.A.; Islam, S.; Aziz, M.A.; Ahammad, A.J.S. Mechanistic insight into hydrothermally prepared molybdenum-based electrocatalyst for overall water splitting. *Electrochim. Acta* **2023**, *445*, 142050. [CrossRef]

28. Ma, G.Y.; Ye, J.T.; Qin, M.Y.; Sun, T.Y.; Tan, W.X.; Fan, Z.H.; Huang, L.F.; Xin, X. Mn-doped NiCoP nanopin arrays as high-performance bifunctional electrocatalysts for sustainable hydrogen production via overall water splitting. *Nano Energy* **2023**, *115*, 108679. [CrossRef]
29. Gu, R.Z.; Zhou, T.; Wang, Z.H.; Chen, Z.; Tao, J.W.; Fan, Z.W.; Guo, L.Y.; Liu, Y.S. Mo/P doped NiFeSe as bifunctional electrocatalysts for overall water splitting. *Int. J. Hydrogen Energy* **2023**, *48*, 27642–27651. [CrossRef]
30. Gao, Y.H.; He, W.J.; Cao, D.; Wang, F.Q.; Li, Y.; Hao, Q.Y.; Liu, C.C.; Liu, H. Mo-Doped Ni_3S_2 Nanosheet Arrays for Overall Water Splitting. *ACS Appl. Nano Mater.* **2023**, *6*, 6066–6075. [CrossRef]
31. You, T.; Deng, K.; Liu, P.; Lv, X.B.; Tian, W.; Li, H.J.; Ji, J.Y. Synergism of NiFe layered double hydroxides/phosphides and Co-NC nanorods array for efficient electrocatalytic water splitting. *Chem. Eng. J.* **2023**, *470*, 144348. [CrossRef]
32. Ning, S.L.; Wu, Q.K.; Zhu, Y.G.; Liu, S.L.; Zhou, W.; Mi, L.; Zhou, K.; Zhao, D.K.; Zhang, X.Y.; Wang, N. N-doped carbon nanowire array confined cobalt phosphides as efficient bifunctional electrocatalysts for water splitting. *Inorg. Chem. Front.* **2023**, *10*, 2145–2153. [CrossRef]
33. Xie, Y.H.; Feng, Y.M.; Jin, S.Y.; Li, C.; Li, C.S.; Sun, Y.; Luo, F.; Yang, Z.H. Nickel-doped iridium echinus-like nanosheets for stable acidic water splitting. *Chem. Commun.* **2023**, *59*, 8404–8407. [CrossRef] [PubMed]
34. Xia, W.; Luan, X.X.; Zhang, W.; Wu, D.F. Sulfuration of hierarchical Mn-doped NiCo LDH heterostructures as efficient electrocatalyst for overall water splitting. *Int. J. Hydrogen Energy* **2023**, *48*, 27631–27641. [CrossRef]
35. Ge, J.M.; Diao, S.T.; Jin, J.X.; Wang, Y.P.; Zhao, X.H.; Zhang, F.Z.; Lei, X.D. NiFeCu phosphides with surface reconstruction via the topotactic transformation of layered double hydroxides for overall water splitting. *Inorg. Chem. Front.* **2023**, *10*, 3515–3524. [CrossRef]
36. Wang, Y.F.; Ye, Q.; Lin, L.; Zhao, Y.X.; Cheng, Y.L. NiFeRu/C and Ru, Fe-Ni_5P_4/C as complementary electrocatalyst for highly efficient overall water splitting. *J. Colloid Interface Sci.* **2023**, *651*, 1008–1019. [CrossRef]
37. Karmakar, A.; Krishnan, A.V.; Jayan, R.; Madhu, R.; Islam, M.M.; Kundu, S. Structural modulation of low-valent iron in LDH-derived Ni_3Se_4 nanosheets: A breakthrough electrocatalyst for the overall water splitting reaction. *J. Mater. Chem. A* **2023**, *11*, 10684–10698. [CrossRef]
38. Xu, X.M.; Pan, Y.L.; Ge, L.; Chen, Y.B.; Mao, X.; Guan, D.Q.; Li, M.R.; Zhong, Y.J.; Hu, Z.W.; Peterson, V.K.; et al. High-Performance Perovskite Composite Electrocatalysts Enabled by Controllable Interface Engineering. *Small* **2021**, *17*, 2101573. [CrossRef]
39. Wang, H.M.; Feng, T.; Wang, L.C.; Hao, W.J. Quantum dot-doped CeO_x–NiB with modulated electron density as a highly efficient bifunctional electrocatalyst for water splitting. *Nanoscale* **2023**, *15*, 6321–6332. [CrossRef]
40. Xu, X.M.; Yang, K.F.; She, J.Z.; Zhai, Y.S.; Zhou, M.H.; Deng, R.H.; Li, Z.M.; Cai, H. Self-supporting and bifunctional Cu-based electrocatalysts with porous structures as superior working electrodes for alkaline water splitting. *CrystEngComm* **2023**, *25*, 1276–1285. [CrossRef]

C Journal of Carbon Research

MDPI

Article

Exploring the Impact of DAHP Impregnation on Activated Carbon Fibers for Efficient Charge Storage and Selective O_2 Reduction to Peroxide

Nemanja Gavrilov [1], Stefan Breitenbach [2,3], Christoph Unterweger [2], Christian Fürst [2] and Igor A. Pašti [1,*]

1 University of Belgrade—Faculty of Physical Chemistry, Studentski trg 12-16, 11158 Belgrade, Serbia; gavrilov@ffh.bg.ac.rs
2 Wood K Plus—Kompetenzzentrum Holz GmbH, Altenberger Strasse 69, 4040 Linz, Austria; s.breitenbach@wood-kplus.at (S.B.); c.unterweger@wood-kplus.at (C.U.); c.fuerst@wood-kplus.at (C.F.)
3 Institute of Chemical Technology of Inorganic Materials (TIM), Johannes Kepler University Linz, Altenberger Strasse 69, 4040 Linz, Austria
* Correspondence: igor@ffh.bg.ac.rs

Abstract: Understanding the properties and behavior of carbon materials is of paramount importance in the pursuit of sustainable energy solutions and technological advancements. As versatile and abundant resources, carbon materials play a central role in various energy conversion and storage applications, making them essential components in the transition toward a greener and more efficient future. This study explores the impact of diammonium hydrogen phosphate (DAHP) impregnation on activated carbon fibers (ACFs) for efficient energy storage and conversion applications. The viscose fibers were impregnated with varying DAHP concentrations, followed by carbonization and activation processes. The capacitance measurements were conducted in 6 mol dm^{-3} KOH, 0.5 mol dm^{-3} H_2SO_4, and 2 mol dm^{-3} KNO_3 solutions, while the oxygen reduction reaction (ORR) measurements were performed in O_2-saturated 0.1 mol dm^{-3} KOH solution. We find that the presented materials display specific capacitances up to 160 F g^{-1} when the DAHP concentration is in the range of 1.0 to 2.5%. Moreover, for the samples with lower DAHP concentrations, highly selective O_2 reduction to peroxide was achieved while maintaining low ORR onset potentials. Thus, by impregnating viscose fibers with DAHP, it is possible to tune their electrochemical properties while increasing the yield, enabling the more sustainable and energy-efficient synthesis of advanced materials for energy conversion applications.

Keywords: activated carbon fibers; energy conversion; energy storage; oxygen reduction reaction

1. Introduction

Carbon materials play a pivotal role in the realm of energy conversion and storage owing to their exceptional properties, widespread availability, and versatility [1]. Their significance is underscored by their usage across various applications, including batteries [2,3], supercapacitors [4,5], fuel cells [6], solar cells [7], carbon capture [8,9], and hydrogen storage systems [10,11].

Some common properties of carbon materials are of high importance for different electrochemical applications. These include a high specific surface area, large porosity, and high conductivity [12]. However, the specific links between different properties and electrochemical performance have been elusive for a long time [13] and still remain a significant challenge for the research community. Irrespective of the complexity of such interlinks, carbon materials are multifunctional materials, and a single carbon material can find many different applications in electrochemical energy conversion and storage. The most direct possible application is in electrochemical capacitors. As there are no electrochemical reactions during the charging/discharging of an electrochemical capacitor, such systems can deliver and store energy very fast, having higher power densities than batteries [14,15]. Thus, electrochemical capacitors can be used to complement battery systems in cases where short power bursts are needed. Many different types of carbon materials have found applications in electrochemical capacitors, like carbon nanotubes [16,17], graphene [17,18], carbon foam [19–21], and others. However, in many cases, the rationale for the use of complex carbon nanostructures is not well-justified. Most carbon materials, independent of their dimensionality, have capacitances up to 200–250 F g^{-1}. Thus, developing carbon materials using cheap and environmentally friendly sources using routes that do not require expensive and harmful chemicals should be the priority. In this sense, biomass-derived carbons present a particularly interesting class of carbon materials due to their structural diversity, renewability, wide availability, environmental friendliness, and low cost [22].

Carbon materials have found their place in electrocatalysis as well. Focusing particularly on fuel cell technologies, carbon materials have proven to be indispensable so far [23–25] for noble metal-based catalysts used on both the anode and cathode sides of low-temperature fuel cells. Focusing particularly on the cathode, where the oxygen reduction reaction (ORR) takes place, carbon materials are not very active ORR catalysts in acidic media but are rather good catalysts in alkaline solutions [26]. Moreover, the ORR can be boosted by introducing heteroatoms, such as nitrogen, boron, or phosphorus, into the carbon structure [27,28]. The introduction of heteroatoms can increase the activity and affect the selectivity of carbons in ORRs. Namely, the ORR is a complex electrochemical reaction that can proceed via a two-electron or four-electron pathway. In the first case, the reaction product is hydrogen peroxide (OH_2^- in alkaline media), while in the second case, it is water. The two-electron pathway is not desirable for fuel cell technologies as it leads to a lower power output (due to a lower number of electrons consumed per O_2) and possible damage to the system, particularly the membrane, by corrosive peroxide. However, this pathway can be utilized for the synthesis of hydrogen peroxide.

Hydrogen peroxide is one of the most frequently used chemicals in the world and has a growing billion-dollar market size [29]. Currently, its classical synthesis utilizes palladium-based catalysts [30], which are expensive and unsuitable for on-site peroxide generation. On the other hand, the electrochemical synthesis of hydrogen peroxide holds significant importance in the field of sustainable chemical production and environmental remediation, offering an alternative to traditional energy-intensive anthraquinone autooxidation processes [31,32]. Electrochemical synthesis offers a cleaner, more efficient, and selective route to H_2O_2 production [33]. This method involves the reduction of oxygen at the cathode, generating H_2O_2 as a primary product. The process can be fine-tuned by controlling parameters such as electrode material, electrolyte composition, and applied potential. This versatility enables the customization of the electrochemical process for various applications, from large-scale industrial production to on-site generation. In particular, electrochemical synthesis offers the advantage of directly producing H_2O_2 at the point of use, minimizing serious transportation, storage, and safety concerns associated with concentrated H_2O_2 solutions [34]. Nevertheless, the first step is identifying the materials with desired ORR selectivity.

To boost the electrochemical synthesis of hydrogen peroxide, catalysts favoring selective 2e-reduction are needed. Different classes of materials were tested for the electrochemical generation of H_2O_2. These include noble metals such as Ag and Au [35,36], oxide materials [37], and particularly, carbon-based materials [38–41]. Carbon materials are particularly suitable for peroxide synthesis as they possess a high chemical stability and can be tuned to favor the 2e-pathway. In this way, the electrochemical route to H_2O_2 synthesis can be stimulated, offering a greener alternative by reducing the environmental footprint compared to traditional H_2O_2 synthesis. Further benefits could be obtained by coupling electrochemical H_2O_2 synthesis with renewable energy sources like solar or wind power to enhance its sustainability further.

In this study, we analyze a series of well-characterized carbon materials in terms of their capacitive and electrocatalytic (ORR) performance. Specifically, we investigated the effects of the synthesis conditions of diammonium hydrogen phosphate (DAHP)-impregnated viscose-based activated carbon fibers (ACFs) on their electrochemical performance. The systematic control of the amount of DAHP allowed for the continuous modification of the properties of the studied carbon materials, while increasing their yield significantly when compared to ACFs obtained without impregnation [42]. Considering the multifunctional nature of carbon materials, we have analyzed the capacitive and electrocatalytic properties of the presented series of materials. The capacitance was evaluated in alkaline (6 mol dm^{-3} KOH), pH-neutral (2 mol dm^{-3} KNO_3), and acidic solutions (0.5 mol dm^{-3} H_2SO_4). Moreover, the ORR was evaluated in an alkaline solution (0.1 mol dm^{-3} KOH) using the rotating disk electrode technique. We have discussed the role of the materials' properties, governed by specific synthetic conditions, on the electrochemical behavior of the studied ACFs. We find that in the analyzed series of materials, the best performance in terms of the capacitive response and selective O_2 reduction to hydrogen peroxide is obtained with intermediate concentrations of DAHP as the impregnation agent, while the specific links between the materials' properties and performance remained elusive.

2. Materials and Methods

2.1. Materials Synthesis and Characterization

The materials' synthesis (previously described in detail in [42]) involved drying viscose fibers (1.7 dtex, 38 mm) at 90 °C for 24 h, followed by impregnation in DAHP solutions of varying concentrations for 15 min. Carbonization was performed at 850 °C (heating rate 1 °C min^{-1}, with an isothermal step for 20 min) under a nitrogen atmosphere, and subsequent activation occurred at 870 °C for 165 min in a CO_2 flow. The ACF samples are denoted as DAHP-X, where X represents the concentration of DAHP used. The synthesis steps are summarized in Scheme S1 (Supplementary Information).

The materials' characterization involved examining the ACF samples' morphology and elemental composition using a scanning electron microscope with energy-dispersive X-ray spectroscopy. Textural properties and specific surface area were determined via N_2 isothermal adsorption using the BET method and NLDFT, respectively [43,44]. Further details can be found in the previous works, in particular Refs. [42,45].

2.2. Capacitance Measurements

For the capacitance measurements, thin film electrodes were prepared through drop-casting of carbon ink. To begin, 5 mg of the desired carbon material was dispersed in a solution consisting of 400 µL of ethanol, 590 µL of deionized water, and 10 µL of a 0.5 wt% Nafion solution in ethanol. The mixture was then homogenized using ultrasound for 30 min to ensure uniformity. Next, films were formed by evaporating the solvent under a mild nitrogen gas stream from a 10 µL droplet of the carbon ink adhered to a glassy carbon (GC) disk electrode with a geometric surface area of 0.196 cm^2. The specific mass loading was set at 250 µg cm^{-2}, which was crucial for the subsequent calculation of the materials' capacitance. Electrochemical measurements were performed in triplicate, starting with the electrode preparation and according to the descriptions provided below.

Electrochemical tests were conducted in a standard three-electrode setup, with a graphite rod serving as the counter electrode and a saturated calomel electrode (SCE) as a reference electrode. Cyclic voltammetry was employed to assess the capacitance of the tested materials in various solutions, including 6 mol dm^{-3} aqueous KOH, 0.5 mol dm^{-3} aqueous H_2SO_4, and 2 mol dm^{-3} aqueous KNO_3 solutions. The Ivium VO1107 potentiostat/galvanostat was utilized for the measurements.

The gravimetric capacitances (C, in F g^{-1}) were calculated using the following formula:

$$C = \frac{Q}{2 \cdot \Delta V \cdot m} = \frac{\int idt}{2 \cdot \Delta V \cdot m} \quad (1)$$

where Q represents the charge passed, obtained by integrating the positive and negative sweep in cyclic voltammograms. ΔV indicates the explored potential window, and m denotes the mass of the active material deposited on the surface of the GC electrode.

Before each experiment, the electrolyte was purged with nitrogen for 15 min to eliminate any potential contaminants. During the measurement, a gentle flow of nitrogen was maintained just below the electrolyte surface to ensure proper conditions. All measurements were performed at room temperature, precisely (25.0 ± 0.5) °C.

2.3. ORR Measurements

The preparation of the electrodes for ORR measurements followed the same procedure as for the capacitance measurements. Briefly, the catalyst thin films were formed by evaporating the solvent under a mild nitrogen gas stream from a 10 μL droplet of the carbon catalyst ink drop-casted to a GC disk electrode with a geometric surface area of 0.196 cm^2. The specific mass loading remained at 250 μg cm^{-2}, as used in the capacitance measurements, ensuring consistency in the experimental setup. To investigate the oxygen reduction reaction (ORR) kinetics, the rotating disc electrode (RDE) technique was employed. The measurements were conducted in an O_2-saturated 0.1 mol dm^{-3} KOH solution, which allowed for the accurate evaluation of the electrocatalytic activity of the carbon materials during the ORR process.

During the experiments, the electrolyte was carefully saturated with oxygen to create an oxygen-rich environment, promoting the ORR at the electrode surface. The RDE technique enabled precise control over the rotation rate of the electrode, ensuring a well-defined hydrodynamic condition for the measurements. Further data processing is elaborated on in Section 3: Results.

3. Results

The materials used in the present work have been extensively characterized and described in our previous works [42,45]. In the aforementioned works, one can find the scanning electron microscopy analysis, EDX data with mapping, specific surface area and pore size distribution measurements, and Raman and FTIR spectra of the described materials. For completeness, here, we present a brief summary of the materials' properties, necessary to discuss the materials' performances, while the representative results of the physical and chemical characterization can be found in Figure S1 (Supplementary Information).

Utilizing SEM, a consistent morphology across all the prepared activated carbon fiber samples was observed. This mirrors that of the precursor viscose fibers with an approximate 30% shrinkage, consistent with the prior findings of Breitenbach et al. [46]. During the milling of the samples, some ACFs remained intact, with lengths of up to 200 μm and a diameter of around 8 μm, while some ACFs fractured into smaller fragments.

Diverging from the uniform morphology, the chemical composition of ACFs demonstrated susceptibility to the loading of DAHP during impregnation (Table 1). As the amount of DAHP increased, the P content and, overall, the O content in the ACFs increased. Thus, by changing the amount of DAHP used in the impregnation step, it was possible to control the amount of P incorporated into the ACFs effectively. On the other hand, the increase in the oxygen content makes the surface more oxophilic, making the ACFs easier to disperse in water and obtain uniform catalyst films on the electrode.

Regarding the specific surface area (SSA) analysis, the sample that was not impregnated with DAHP showed a high SSA of nearly 2000 m^2 g^{-1}. Then, the initial P-containing sample exhibited a considerable SSA decrease compared to the ACF sample without DAHP impregnation. However, the introduction of higher DAHP concentrations resulted in a subsequent SSA and total pore volume (V_{tot}) increase. Interestingly, the DAHP-2 sample deviated from this trend, showing a noticeably higher SSA compared to the samples with similar concentrations of DAHP used for the impregnation (1.5% and 2.5%). Moreover, the EDX findings showcased a heightened P fraction within its structure. Another interesting observation is that the pores shift towards larger radii as the DAHP concentration increases.

Namely, when the pore volume for the pores with a diameter under 1 nm is considered, there is clearly an overall decreasing trend in the V_{1nm} as the DAHP concentration increases (Table 1). At the moment, we are not sure about the origin of these deviations from the overall trends when the DAHP concentration is between 1.5 and 2.5%. However, tuning the physical and chemical properties of the studied ACFs in a relatively wide and controllable way can help not only identify the most suitable materials for the desired applications but also draw some general conclusions regarding the impact of the materials' properties on their electrochemical behavior.

Table 1. Adsorbents properties—pore volume integrated up 1 nm, total pore volume (integrated up to d_p = 35 nm, V_{tot}), surface area (S_{tot}), and elemental content determined using energy-dispersive X-ray spectroscopy. The notation of samples is DAHP-X, where X represents the amount of DAHP used for the impregnation step. Reproduced from [42]. This work is published with Open Access under Creative Commons License.

X=		0.00%	0.25%	0.50%	1.00%	1.50%	2.00%	2.50%	5.00%	7.50%	10%
$V_{1nm}/cm^3\ g^{-1}$		0.348	0.335	0.339	0.343	0.297	0.271	0.274	0.198	0.137	0.160
$V_{tot}\ cm^3\ g^{-1}$		0.757	0.383	0.472	0.774	0.770	1.094	0.833	1.291	1.681	1.322
$S_{tot}/m^2\ g^{-1}$		1932	1016	1250	2037	2002	2556	2018	2718	2763	2718
Elemental Content	C/at.%	92.4 ± 2.1	91.6 ± 3.5	93.6 ± 2.3	87.9 ± 1.3	93.9 ± 1.9	91.3 ± 2.0	87.9 ± 2.0	85.6 ± 2.5	82.2 ± 2.2	77.8 ± 5.0
	O/at.%	7.6 ± 2.0	8.4 ± 3.5	6.2 ± 2.2	12 ± 1.3	5.9 ± 1.9	7.7 ± 2.3	11.4 ± 2.2	13.1 ± 2.7	16.1 ± 2.3	19.7 ± 5.3
	P/at.%	0	0.02 ± 0.02	0.18 ± 0.10	0.11 ± 0.04	0.28 ± 0.07	0.91 ± 0.29	0.65 ± 0.30	1.32 ± 0.34	1.78 ± 0.13	1.9 ± 0.64

First, we present the results of the capacitance measurements. Figure 1 (top row) shows the capacitive responses of the DAHP-2.0 sample in three different electrolytes. As we have used the SCE electrode as a reference electrode, the potential window in which cycling was performed is shifted to higher potentials as the pH of the electrolytic solution decreases. It is clear that the capacitive response is nearly rectangular in shape, without significant features like voltammetry peaks that would indicate pseudo-faradaic or faradic behavior. The overall capacitive response considering the different electrolytic solutions increases in the following order: $KNO_3 \approx< H_2SO_4 < KOH$. Such a trend is clearly seen from the comparison of the specific capacitances measured for the entire series at a common potential sweep rate of 100 mV s^{-1} (Figure 1, bottom row). Overall, considering the capacitances measured at 100 mV s^{-1}, for the entire series, the values range between 35 and 160 F g^{-1}, demonstrating a great variety of the capacitive properties of the studied materials. This result is not overly surprising, considering that the chemical composition and the textural properties of the studied carbon materials also vary to a significant extent. However, while the properties of the studied carbons change progressively with the amount of DAHP used in the synthesis, the capacitive properties do not show such a trend, and the capacitance is maximized for the samples prepared with a relatively low amount of impregnating agent (1.0% to 2.5%).

Table 2 summarizes the measured capacitances of all the samples for potential sweep rates from 5 to 400 mV s^{-1} in 6 mol dm^{-3} KOH solution. For the KNO_3 and H_2SO_4 solutions, the data are given in the Supplementary Information, Table S1 and S2. Figure 2 presents the overall trends in the measured capacitances.

Table 2. Measured capacitances (F g^{-1}) in 6 mol dm^{-3} KOH solution at different scan rates.

Material	5 mV s^{-1}	10 mV s^{-1}	20 mV s^{-1}	50 mV s^{-1}	100 mV s^{-1}	200 mV s^{-1}	400 mV s^{-1}
0.0% DAHP	84.5 ± 6.8	87.6 ± 6.1	89.4 ± 6.3	87.2 ± 5.2	75.5 ± 4.5	69.1 ± 4.1	59.5 ± 4.2
0.25% DAHP	94.6 ± 7.6	88.4 ± 4.4	82.2 ± 4.1	70.9 ± 4.3	56.6 ± 3.4	46.0 ± 2.8	36.6 ± 2.9
0.5% DAHP	87.8 ± 7.9	78.1 ± 6.2	75.8 ± 4.5	74.9 ± 3.0	62.6 ± 2.5	57.7 ± 2.3	51.9 ± 2.6
1.0% DAHP	133 ± 7	149 ± 10	168.2 ± 6.7	174 ± 11	159 ± 8	151 ± 11	138 ± 11

Table 2. *Cont.*

Material	5 mV s^{-1}	10 mV s^{-1}	20 mV s^{-1}	50 mV s^{-1}	100 mV s^{-1}	200 mV s^{-1}	400 mV s^{-1}
1.5% DAHP	192 ± 8	137.3 ± 6.9	144 ± 7	148.0 ± 5.9	135.0 ± 8.1	128.2 ± 6.4	118.1 ± 9.4
2.0% DAHP	125 ± 11	146 ± 6	168 ± 7	173 ± 7	156.2 ± 7.8	150 ± 14	138 ± 11
2.5% DAHP	141.9 ± 8.5	152.2 ± 6.1	166 ± 10	167 ± 10	153.0 ± 6.1	142.8 ± 5.7	128.9 ± 6.4
5.0% DAHP	92.5 ± 7.4	93.0 ± 7.4	113.0 ± 4.5	126.0 ± 6.3	112.7 ± 7.9	112.4 ± 7.9	107.4 ± 7.5
7.5% DAHP	91.4 ± 3.7	77.8 ± 6.2	84.2 ± 5.9	88.9 ± 5.3	78.3 ± 4.7	75.2 ± 6.0	69.7 ± 4.9
10.0% DAHP	124.7 ± 7.5	107.3 ± 6.4	110.7 ± 6.6	115.4 ± 8.1	99.8 ± 4.0	96.9 ± 8.7	91.5 ± 6.4

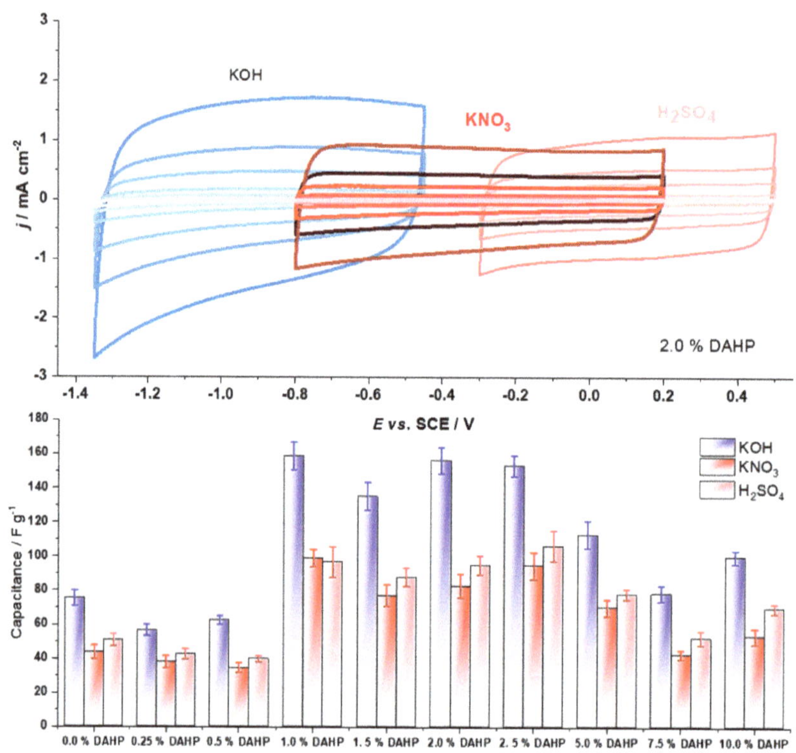

Figure 1. Top row—cyclic voltammograms of 2.0% DAHP sample at different scan rates and electrolytes at a common potential sweep rate of 100 mV s^{-1}; **bottom** row—measured capacitances at a potential scan rate of 100 mV s^{-1} in different electrolytes (quiescent and N_2-purged).

As can be seen from the presented data, Figure 1 (bottom row), and Figure 2, the samples in the middle of the series with a DAHP concentration ranging from 1.0 to 2.5% stand out as the ones with the highest capacitive responses in all three electrolytes. This finding immediately draws to the conclusion that the SSA is not the primary factor that determines the capacitive response, as the highest SSA is reached for the materials impregnated with high concentrations of DAHP.

Next, we turn to the ORR activity measurements. First, we compare the background curves, i.e., the capacitive responses in 0.1 mol dm^{-3} KOH (Figure 3, left). The baselines suggest that the used electrolyte's electrochemically accessible surface differs from sample to sample and follows the trends in the measured capacitances in the concentrated KOH solution. Upon introducing O_2 in the solution, the ORR current was measured (Figure 3, middle) at different electrode rotation rates (Figure 3, right). The onset potentials for some of the materials were found to be rather high, above −0.2 V vs. SCE (i.e., above 0.8 V vs. RHE), with the sample denoted as 2.5% DAHP having the highest activity considering the current response. Further, to precisely evaluate the activity and selectivity of the studied materials, we have used Koutecky–Levich (K–L) analysis [47].

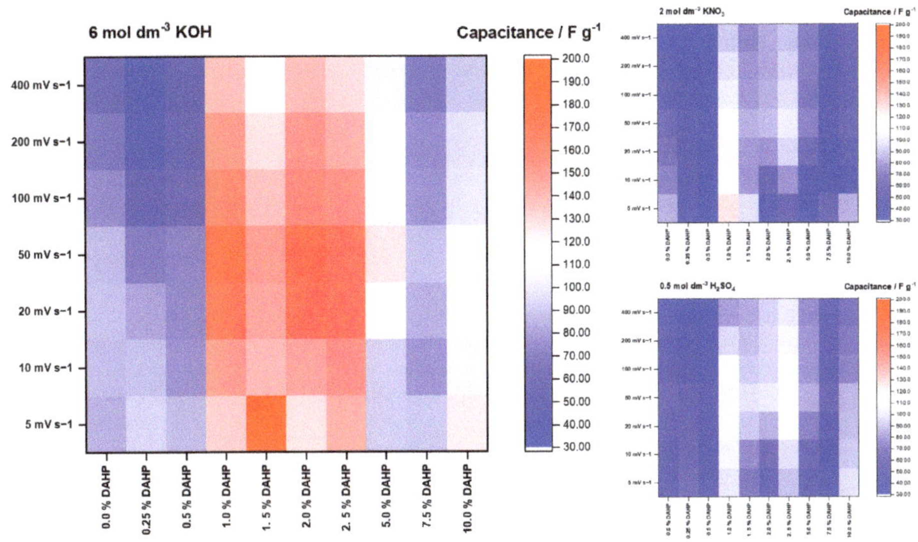

Figure 2. Heatmaps for measured capacitances in three different solutions, as indicated in the plots. The potential sweep rate was between 5 and 400 mV s^{-1}. The color map and the scale are the same for all three electrolytes, which clearly shows that the capacitances are the highest in the 6 mol dm^{-3} KOH solution.

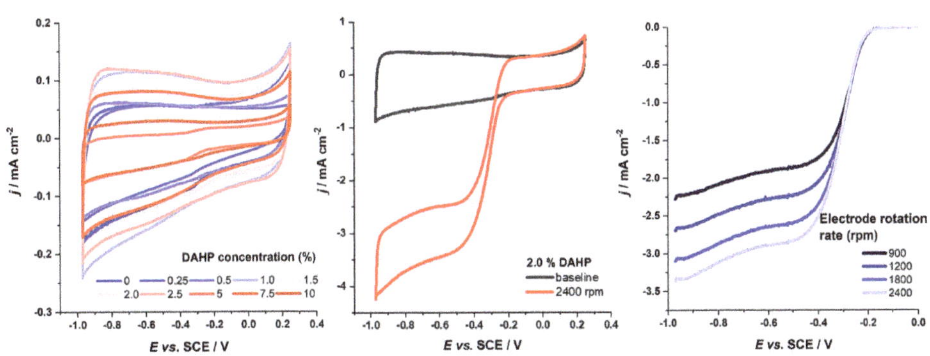

Figure 3. **Left**—baselines for the investigated materials in de-aerated 0.1 mol dm^{-3} KOH, potential sweep rate 20 mV s^{-1}; **middle**—raw ORR current (at the electrode rotation rate of 2400 rpm) and the background current of the DAHP-2.0 sample; **right**—background-corrected ORR currents of 2.5% DAHP sample, measured at different electrode rotation rates (in rpm, indicated in the figure).

In the K–L analysis, the measured current densities (j) can be presented as (see Figure S2, Supplementary Information

$$\frac{1}{j} = \frac{1}{j_k} + \frac{1}{j_d} \quad (2)$$

where j_k and j_d are the kinetic current and limiting diffusion current, respectively. The kinetic current depends primarily on the electrode potential, while the diffusion current depends on the number of electrons transferred per O_2 molecule (n) and the electrode rotation rate (ω) as $j_d = n \times B \times \omega^{1/2}$. B is a constant that assembles other parameters such as the O_2 concentration, kinematic viscosity of the solution, and other relevant parameters that were constant during our measurements [47]. Once we extracted the kinetic current densities, we normalized them to the mass of the ACFs deposited on the electrode (by dividing j_k with the catalyst loading), thus obtaining mass-specific activities at different electrode potentials (Table S3, Supplementary Information). The calculated numbers of electrons consumed per O_2 molecule are summarized in the Supplementary Information, Table S4.

The number of electrons and the kinetic current are mutually interconnected, as determined in the K–L analysis from the slope and the intercept of the K–L plot (Equation (2)). Figure 4 (top row) summarizes the mass activities (in A g^{-1}) at two different electrode potentials (−0.4 and −0.6 V vs. SCE), while Figure 4 (bottom row) presents the dependence of the number of electrons per O_2 molecule on the electrode potential for the entire series of ACFs. We have not observed a particular trend along the series considering the mass activities, while the samples with lower concentrations of DAHP (thus P and O, Table 1) show a somewhat lower number of electrons transferred per O_2 molecule, thus show a better selectivity for peroxide formation. Nevertheless, the samples in the middle of the series, which exhibited the highest capacitances in alkaline solution (Figure 2), do show the highest ORR mass activities. However, the overall trends in the capacitances and ORR mass activities are not the same.

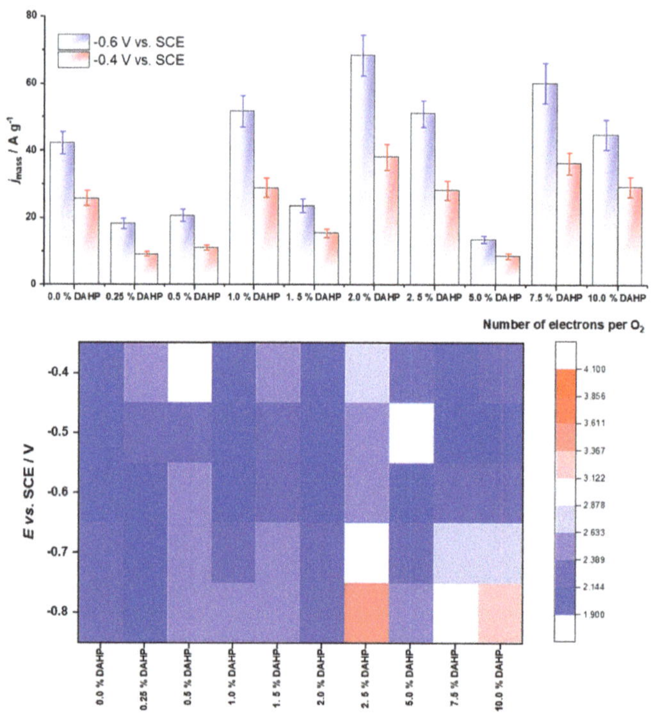

Figure 4. **Top** row—ORR mass activities of investigated samples at −0.4 and −0.6 V vs. SCE, calculated from the anodic scan; **bottom** row—number of electrons consumed per O_2 molecule, in the potential range −0.4 to −0.8 V vs. SCE.

4. Discussion

Overall, in the analysis of the measured capacitance, we observed some irregular behavior concerning its dependence on the scan rate, particularly a decrease in the capacitance at low potential sweep rates. Notably, this effect is no longer evident at high potential scan rates. Based on our findings, we can conclude that the capacitance is maximized for DAHP concentrations ranging between 1% and 2.5%. Furthermore, we noticed a consistent trend in the capacitance values across different electrolytes, with KOH exhibiting the highest capacitance. The cyclic voltammograms displayed a well-defined capacitive behavior for most of the tested materials, further supporting the observed high capacitance in alkaline media. Our results align with our previous experiences involving various carbon materials, where high capacitances were consistently observed to be the highest in an alkaline environment [48–51]. In the analysis of the ORR activity, we observed variations in the catalytic activity along the series but without any particular trend that can be outlined. Notably, among the tested materials, the sample denoted as DAHP-2.0 exhibited the highest ORR activity, measured as mass activity. When considering the entire series of materials, we did not observe any prominent plateau with increasing the DAHP content, but rather variations in electrochemical response, which are more prominent in the case of the ORR activity (Figure 4). In the case of the capacitive response (Figure 1) one could hypothesize that the DAHP content above 2.5% starts to have a detrimental effect on the capacitance, either because the pores become too large or the surface chemistry is altered in undesirable ways. However, such a hypothesis cannot be valid for the case of the ORR activity as the activity oscillates along the series. Next, we discuss the connection between the materials' properties and their electrochemical behavior.

In the analysis of the properties–performance link, we have investigated the correlation of the different materials' properties, as summarized in Table 1, and the results of the electrochemical investigation. Strikingly, while the materials' properties, like the O content, SSA, and pore volumes, are mutually interconnected, their link with the materials' performance is missing (Figure 5). The consistency between the measured capacitances in different electrolytes was observed, showing that the overall trends in the capacitances are preserved in different electrolytes. However, we have failed to find any properties that can be significantly correlated to the ORR activity and measured capacitances. Moreover, the attempt to use the linear combinations of materials' properties through the multiple linear regression approach was also not successful. Likely, there is a very complex property–performance dependence that cannot be outlined in a simple way. This urges further research in an attempt to reveal the deep links between the materials' structure and electrochemical performance.

Nevertheless, the samples with relatively low amounts of DAHP (1.0 to 2.5%) stand out as materials with a high capacitance and also high ORR activity. Moreover, the same materials displayed a high selectivity of O_2 reduction to peroxide. The DAHP-2.0 sample displayed a high capacitance in alkaline solutions (up to approx. 170 F g^{-1}) and up to 100 F g^{-1} in acidic solutions. The same sample showed mass activity of more than 60 A g^{-1} at high ORR overpotentials (−0.8 V vs. SCE) and a very good selectivity of O_2 reduction to peroxide. Namely, the calculated number of electrons consumed per O_2 was between 2.06 and 2.35 in the potential range from −0.4 to −0.8 V vs. SCE. Notably, there were a few more samples with relatively high mass activities and good selectivity for peroxide formation, especially at higher potentials (lower ORR overpotentials). These include the DAHP-0, DAHP-1.0, and DAHP-7.5 samples. On the other hand, the DAHP-2.5 sample showed the highest n, setting itself as the most active material for O_2 reduction to water (hydroxyl ions). Considering that the ACF yields are significantly improved upon the impregnation with DAHP [52,53] and that the ORR onset potentials are quite high (above 0.8 V vs. RHE, that is, above −0.2 V vs. SCE), the described ACF synthesis seems to be a viable strategy to obtain highly efficient ORR catalysts with a potential to be used in the electrochemical synthesis of hydrogen peroxide. However, the present results should be taken with care regarding the actual use of the studied materials for electrochemical hydrogen peroxide synthesis, and additional work is needed, particularly determining the H_2O_2 yield under relevant synthesis conditions and determining the catalysts' stability under prolonged operation.

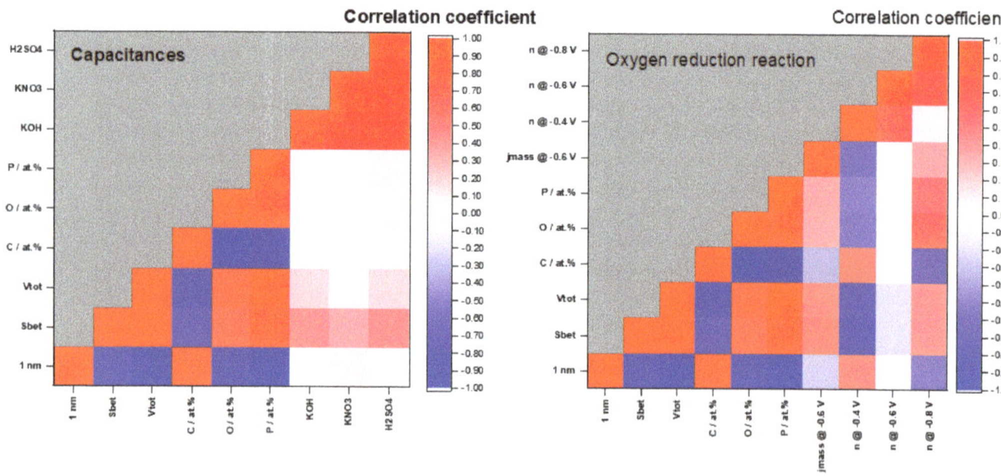

Figure 5. Pearson's correlation coefficient matrix linking capacitance (**left**) and ORR activity and selectivity (**right**) with materials properties.

We note that the capacitive applications of carbon materials usually relate to their electrochemistry in non-aqueous solutions. This is because the energy density of an electrochemical capacitor is not only a linear function of the gravimetric capacitance but also a square function of the voltage window. In this way, by making a wider window of the electrochemical stability of capacitive material and the electrolyte, one can significantly increase the energy density of the capacitor. Moreover, the electrochemical capacitor market is predominantly populated with devices containing non-aqueous electrolytes (usually with an upper voltage limit of approximately 2.7 V), but the majority of fundamental studies are performed in aqueous electrolytes. Part of the presented series of materials has already been analyzed in terms of the capacitance in a non-aqueous electrolyte [53]. In that study, the capacitances up to 139 F g^{-1} were measured; thus, they were somewhat lower compared to the values measured here in KOH solution but higher compared to the ones measured in the H_2SO_4 and KNO_3 solutions. Moreover, just like the present study, the measured capacitances were not possible to uniquely link to some of the specific materials' properties, like the SSA or pore volume. To make the situation more complicated, the maximum capacitance in non-aqueous solutions was observed for the sample with the highest concentration of P, while, here, we see the maximum capacitance for the DAHP-2.0 sample. This particular sample was indicated to stand out by a high SSA and higher P content than one could expect on the basis of the initial DAHP concentration used for impregnation. Thus, not only do the links between the carbon materials' properties and their electrochemical performance remain elusive but also the electrochemical performance in aqueous and non-aqueous solutions does not correlate mutually.

5. Conclusions

In this study, the impact of DAHP impregnation in the production of viscose-derived ACFs for electrochemical applications was investigated. The DAHP concentration significantly influences the capacitance of the ACFs, with an optimum range of 1.0% to 2.5%. Furthermore, the highest capacitance was consistently achieved in the KOH electrolyte, followed by H_2SO_4, and finally KNO_3. The analysis of the ORR activity indicated that some materials exhibited high onset potentials, exceeding −0.2 V vs. SCE. Particularly, the

DAHP-2.0 sample displayed the highest activity among the tested materials, expressed in terms of mass activity. As a particularly important finding, we emphasize the high selectivity of O_2 reduction to peroxide, which can be utilized for the electrochemical synthesis of hydrogen peroxide. Several materials in the investigated series, including the most active, the DAHP-2.0 sample, displayed the almost complete O_2 reduction to peroxide, with the number of transferred electrons per O_2 molecule practically matching two at higher potentials. Considering the fine-tuning and optimization of the materials' properties, the obtained results suggest that the DAHP concentration range from 1.0 to 2.5% should be investigated in more detail. Our attempts to link the materials' properties and their electrochemical performances were unsuccessful, making these links elusive. Moreover, we could not correlate the capacitances measured in aqueous solutions with those measured in non-aqueous electrolytes. This leads to the conclusion that specific interactions governing high capacitive responses in aqueous and non-aqueous electrolytes are different. While the present study did not allow for the deep link between the physicochemical properties of materials and their electrochemical behavior, it clearly emphasizes the positive role of DAHP impregnation on tuning the ACFs' electrochemical behavior in terms of the capacitive response and ORR activity and selectivity. Combining these findings with a significant increase in ACF yields when the DAHP impregnation step is used, it is clear that the demonstrated approach could lead to more energy-efficient and sustainable routes for producing advanced materials with desirable electrochemical properties.

Supplementary Materials: The following supporting information can be downloaded at: https://www.mdpi.com/article/10.3390/c9040105/s1, Scheme S1: The block synthesis workflow for the DAHP-X samples; Figure S1: The representative results of the physical and chemical characterization of the studied samples: top row—Raman spectrum and Pore Size Distribution for selected samples, middle row—Scanning Electron Microscopy Analysis, bottom row—EDX mapping for the sample 2% DAHP; Figure S2: Representative Koutecky-Levich plots of studied ACF samples: (a) sample 2.0% DAHP, (b) sample 2.5% DAHP; Table S1: Measured capacitances (F g^{-1}) in H_2SO_4 solution at different scan rates; Table S2: Measured capacitances (F g^{-1}) in KNO_3 solution at different scan rates; Table S3: Calculated ORR mass activities (in A g^{-1}); Table S4: Number of electrons consumed *per* O_2 molecule, calculated at different electrode potentials.

Author Contributions: Conceptualization, N.G., S.B., C.U., C.F. and I.A.P.; formal analysis, N.G., S.B. and I.A.P.; funding acquisition, C.F. and I.A.P.; investigation, N.G. and S.B.; methodology, N.G., S.B., C.U., C.F. and I.A.P.; project administration, C.F.; resources, C.U.; supervision, C.U. and C.F.; validation, S.B. and C.U.; writing—original draft, N.G., S.B. and I.A.P.; writing—review and editing, C.U., C.F. and I.A.P. All authors have read and agreed to the published version of the manuscript.

Funding: S.B., C.U. and C.F. gratefully acknowledge financial support through the COMET Programme (Competence Centers for Excellent Technologies) funded by the Austrian ministries BMK, BMAW, and the federal states of Upper Austria, Lower Austria, and Carinthia, operated by the Austrian Research Promotion Agency (FFG). I.A.P. and N.G. acknowledge the support provided by the Serbian Ministry of Education, Science, and Technological Development (Contract number: 451-03-47/2023-01/200146).

Data Availability Statement: The data are contained within the article.

Conflicts of Interest: The authors declare no conflict of interest.

References

1. Zhang, Y.; Li, C.; Cao, F.; Noori, A.; Mousavi, M.F.; Xia, X. Carbon in Electrochemical Energy. *Mater. Res. Bull.* **2022**, *152*, 111852. [CrossRef]
2. Wu, Z.; Sun, K.; Wang, Z. A Review of the Application of Carbon Materials for Lithium Metal Batteries. *Batteries* **2022**, *8*, 246. [CrossRef]
3. Yang, S.; Cheng, Y.; Xiao, X.; Pang, H. Development and Application of Carbon Fiber in Batteries. *Chem. Eng. J.* **2020**, *384*, 123294. [CrossRef]
4. Wang, Y.; Zhang, L.; Hou, H.; Xu, W.; Duan, G.; He, S.; Liu, K.; Jiang, S. Recent Progress in Carbon-Based Materials for Supercapacitor Electrodes: A Review. *J. Mater. Sci.* **2021**, *56*, 173–200. [CrossRef]

5. Zhai, Z.; Zhang, L.; Du, T.; Ren, B.; Xu, Y.; Wang, S.; Miao, J.; Liu, Z. A Review of Carbon Materials for Supercapacitors. *Mater. Des.* **2022**, *221*, 111017. [CrossRef]
6. Jaleh, B.; Nasrollahzadeh, M.; Eslamipanah, M.; Nasri, A.; Shabanlou, E.; Manwar, N.R.; Zboril, R.; Fornasiero, P.; Gawande, M.B. The Role of Carbon-Based Materials for Fuel Cells Performance. *Carbon* **2022**, *198*, 301–352. [CrossRef]
7. Hadadian, M.; Smått, J.-H.; Correa-Baena, J.-P. The Role of Carbon-Based Materials in Enhancing the Stability of Perovskite Solar Cells. *Energy Environ. Sci.* **2020**, *13*, 1377–1407. [CrossRef]
8. Abd, A.A.; Othman, M.R.; Kim, J. A Review on Application of Activated Carbons for Carbon Dioxide Capture: Present Performance, Preparation, and Surface Modification for Further Improvement. *Environ. Sci. Pollut. Res.* **2021**, *28*, 43329–43364. [CrossRef] [PubMed]
9. Liu, E.; Lu, X.; Wang, D. A Systematic Review of Carbon Capture, Utilization and Storage: Status, Progress and Challenges. *Energies* **2023**, *16*, 2865. [CrossRef]
10. Bader, N.; Ouederni, A. Optimization of Biomass-Based Carbon Materials for Hydrogen Storage. *J. Energy Storage* **2016**, *5*, 77–84. [CrossRef]
11. Mohan, M.; Sharma, V.K.; Kumar, E.A.; Gayathri, V. Hydrogen Storage in Carbon Materials—A Review. *Energy Storage* **2019**, *1*, e35. [CrossRef]
12. Pang, T.; Marken, F.; Zhang, D.; Mattia, D.; Shen, J. Linking Macroscopic Surface Morphology of Activated Carbon Fibres and Electrosorption Performance: An Electrochemical Impedance Spectroscopy and Capacitive Deionization Study. *Appl. Surf. Sci.* **2023**, *609*, 155397. [CrossRef]
13. Pašti, I.A.; Janošević Ležaić, A.; Gavrilov, N.M.; Ćirić-Marjanović, G.; Mentus, S.V. Nanocarbons Derived from Polymers for Electrochemical Energy Conversion and Storage—A Review. *Synth. Met.* **2018**, *246*, 267–281. [CrossRef]
14. Winter, M.; Brodd, R.J. What Are Batteries, Fuel Cells, and Supercapacitors? *Chem. Rev.* **2004**, *104*, 4245–4270. [CrossRef] [PubMed]
15. Rani, J.; Thangavel, R.; Oh, S.-I.; Lee, Y.; Jang, J.-H. An Ultra-High-Energy Density Supercapacitor; Fabrication Based on Thiol-Functionalized Graphene Oxide Scrolls. *Nanomaterials* **2019**, *9*, 148. [CrossRef] [PubMed]
16. Felhősi, I.; Keresztes, Z.; Marek, T.; Pajkossy, T. Properties of Electrochemical Double-Layer Capacitors with Carbon-Nanotubes-on-Carbon-Fiber-Felt Electrodes. *Electrochim. Acta* **2020**, *334*, 135548. [CrossRef]
17. Frackowiak, E.; Jurewicz, K.; Delpeux, S.; Beguin, F. Electrochemical Application of Carbon Nanotubes. In *Low-Dimensional Systems: Theory, Preparation, and Some Applications*; Springer: Dordrecht, The Netherlands, 2003; pp. 305–318.
18. Chen, J.; Li, C.; Shi, G. Graphene Materials for Electrochemical Capacitors. *J. Phys. Chem. Lett.* **2013**, *4*, 1244–1253. [CrossRef]
19. Udayakumar, M.; Tóth, P.; Wiinikka, H.; Malhotra, J.S.; Likozar, B.; Gyergyek, S.; Leskó, A.K.; Thangaraj, R.; Németh, Z. Hierarchical Porous Carbon Foam Electrodes Fabricated from Waste Polyurethane Elastomer Template for Electric Double-Layer Capacitors. *Sci. Rep.* **2022**, *12*, 11786. [CrossRef] [PubMed]
20. Cao, M.; Feng, Y.; Tian, R.; Chen, Q.; Chen, J.; Jia, M.; Yao, J. Free-Standing Porous Carbon Foam as the Ultralight and Flexible Supercapacitor Electrode. *Carbon* **2020**, *161*, 224–230. [CrossRef]
21. Park, H.; Seo, J.; Kim, M.; Baeck, S.-H.; Shim, S.E. Development of a Carbon Foam Supercapacitor Electrode from Resorcinol–Formaldehyde Using a Double Templating Method. *Synth. Met.* **2015**, *199*, 121–127. [CrossRef]
22. Zhang, Y.; Pan, H.; Zhou, Q.; Liu, K.; Ma, W.; Fan, S. Biomass-Derived Carbon for Supercapacitors Electrodes—A Review of Recent Advances. *Inorg. Chem. Commun.* **2023**, *153*, 110768. [CrossRef]
23. Lam, E.; Luong, J.H.T. Carbon Materials as Catalyst Supports and Catalysts in the Transformation of Biomass to Fuels and Chemicals. *ACS Catal.* **2014**, *4*, 3393–3410. [CrossRef]
24. Rodríguez-Reinoso, F.; Sepúlveda-Escribano, A. Carbon as Catalyst Support. In *Carbon Materials for Catalysis*; John Wiley & Sons, Inc.: Hoboken, NJ, USA, 2008; pp. 131–155.
25. Trogadas, P.; Fuller, T.F.; Strasser, P. Carbon as Catalyst and Support for Electrochemical Energy Conversion. *Carbon* **2014**, *75*, 5–42. [CrossRef]
26. Ma, R.; Lin, G.; Zhou, Y.; Liu, Q.; Zhang, T.; Shan, G.; Yang, M.; Wang, J. A Review of Oxygen Reduction Mechanisms for Metal-Free Carbon-Based Electrocatalysts. *NPJ Comput. Mater.* **2019**, *5*, 78. [CrossRef]
27. Taguchi, T.; Gohda, S.; Gotoh, K.; Sato, S.; Yamada, Y. Synthesis of Carbon Materials with Extremely High Pyridinic-Nitrogen Content and Controlled Edges from Aromatic Compounds with Highly Symmetric Skeletons. *Carbon Lett.* **2023**, *33*, 1279–1301. [CrossRef]
28. Pašti, I.A.; Gavrilov, N.M.; Dobrota, A.S.; Momčilović, M.; Stojmenović, M.; Topalov, A.; Stanković, D.M.; Babić, B.; Ćirić-Marjanović, G.; Mentus, S.V. The Effects of a Low-Level Boron, Phosphorus, and Nitrogen Doping on the Oxygen Reduction Activity of Ordered Mesoporous Carbons. *Electrocatalysis* **2015**, *6*, 498–511. [CrossRef]
29. Tuci, G.; Rossin, A.; Zhang, X.; Truong-Phuoc, L.; Berretti, E.; Liu, Y.; Pham-Huu, C.; Ali, S.; Jan, F.; Poggini, L.; et al. Metal-Free Electrocatalysts for the Selective 2 e^- Oxygen Reduction Reaction: A Never-Ending Story? *Chem.—Eur. J.* **2023**, *29*, e202301036. [CrossRef] [PubMed]
30. Campos-Martin, J.M.; Blanco-Brieva, G.; Fierro, J.L.G. Hydrogen Peroxide Synthesis: An Outlook beyond the Anthraquinone Process. *Angew. Chemie Int. Ed.* **2006**, *45*, 6962–6984. [CrossRef] [PubMed]
31. Foller, P.C.; Bombard, R.T. Processes for the Production of Mixtures of Caustic Soda and Hydrogen Peroxide via the Reduction of Oxygen. *J. Appl. Electrochem.* **1995**, *25*, 613–627. [CrossRef]

32. Gopal, R. Electrochemical Synthesis of Hydrogen Peroxide. U.S. Patent 20030019758A1, 30 March 2004.
33. Perry, S.C.; Pangotra, D.; Vieira, L.; Csepei, L.-I.; Sieber, V.; Wang, L.; Ponce de León, C.; Walsh, F.C. Electrochemical Synthesis of Hydrogen Peroxide from Water and Oxygen. *Nat. Rev. Chem.* **2019**, *3*, 442–458. [CrossRef]
34. Yang, S.; Verdaguer-Casadevall, A.; Arnarson, L.; Silvioli, L.; Čolić, V.; Frydendal, R.; Rossmeisl, J.; Chorkendorff, I.; Stephens, I.E.L. Toward the Decentralized Electrochemical Production of H_2O_2: A Focus on the Catalysis. *ACS Catal.* **2018**, *8*, 4064–4081. [CrossRef]
35. Pizzutilo, E.; Kasian, O.; Choi, C.H.; Cherevko, S.; Hutchings, G.J.; Mayrhofer, K.J.J.; Freakley, S.J. Electrocatalytic Synthesis of Hydrogen Peroxide on Au-Pd Nanoparticles: From Fundamentals to Continuous Production. *Chem. Phys. Lett.* **2017**, *683*, 436–442. [CrossRef]
36. Huang, X.; Song, M.; Zhang, J.; Shen, T.; Luo, G.; Wang, D. Recent Advances of Electrocatalyst and Cell Design for Hydrogen Peroxide Production. *Nano-Micro Lett.* **2023**, *15*, 86. [CrossRef] [PubMed]
37. Chen, Z.; Wu, J.; Chen, Z.; Yang, H.; Zou, K.; Zhao, X.; Liang, R.; Dong, X.; Menezes, P.W.; Kang, Z. Entropy Enhanced Perovskite Oxide Ceramic for Efficient Electrochemical Reduction of Oxygen to Hydrogen Peroxide. *Angew. Chem.* **2022**, *134*, e202200086. [CrossRef]
38. Xia, Y.; Zhao, X.; Xia, C.; Wu, Z.-Y.; Zhu, P.; Kim, J.Y.; Bai, X.; Gao, G.; Hu, Y.; Zhong, J.; et al. Highly Active and Selective Oxygen Reduction to H2O2 on Boron-Doped Carbon for High Production Rates. *Nat. Commun.* **2021**, *12*, 4225. [CrossRef] [PubMed]
39. Xia, C.; Xia, Y.; Zhu, P.; Fan, L.; Wang, H. Direct Electrosynthesis of Pure Aqueous H_2O_2 Solutions up to 20% by Weight Using a Solid Electrolyte. *Science* **2019**, *366*, 226–231. [CrossRef]
40. Kim, H.W.; Ross, M.B.; Kornienko, N.; Zhang, L.; Guo, J.; Yang, P.; McCloskey, B.D. Efficient Hydrogen Peroxide Generation Using Reduced Graphene Oxide-Based Oxygen Reduction Electrocatalysts. *Nat. Catal.* **2018**, *1*, 282–290. [CrossRef]
41. An, J.; Feng, Y.; Zhao, Q.; Wang, X.; Liu, J.; Li, N. Electrosynthesis of H_2O_2 through a Two-Electron Oxygen Reduction Reaction by Carbon Based Catalysts: From Mechanism, Catalyst Design to Electrode Fabrication. *Environ. Sci. Ecotechnol.* **2022**, *11*, 100170. [CrossRef]
42. Jocić, A.; Breitenbach, S.; Bajuk-Bogdanović, D.; Pašti, I.A.; Unterweger, C.; Fürst, C.; Lazarević-Pašti, T. Viscose-Derived Activated Carbons Fibers as Highly Efficient Adsorbents for Dimethoate Removal from Water. *Molecules* **2022**, *27*, 1477. [CrossRef]
43. Bardestani, R.; Patience, G.S.; Kaliaguine, S. Experimental Methods in Chemical Engineering: Specific Surface Area and Pore Size Distribution Measurements—BET, BJH, and DFT. *Can. J. Chem. Eng.* **2019**, *97*, 2781–2791. [CrossRef]
44. Kupgan, G.; Liyana-Arachchi, T.P.; Colina, C.M. NLDFT Pore Size Distribution in Amorphous Microporous Materials. *Langmuir* **2017**, *33*, 11138–11145. [CrossRef] [PubMed]
45. Jocić, A.; Breitenbach, S.; Pašti, I.A.; Unterweger, C.; Fürst, C.; Lazarević-Pašti, T. Viscose-Derived Activated Carbons as Adsorbents for Malathion, Dimethoate, and Chlorpyrifos—Screening, Trends, and Analysis. *Environ. Sci. Pollut. Res.* **2022**, *29*, 35138–35149. [CrossRef]
46. Breitenbach, S.; Gavrilov, N.; Pašti, I.; Unterweger, C.; Duchoslav, J.; Stifter, D.; Hassel, A.W.; Fürst, C. Biomass-Derived Carbons as Versatile Materials for Energy-Related Applications: Capacitive Properties vs. Oxygen Reduction Reaction Catalysis. *C* **2021**, *7*, 55. [CrossRef]
47. Bard, A.J.; Faulkner, L.R. *Electrochemical Methods: Fundamentals and Applications*; Wiley: New York, NY, USA, 1980.
48. Milakin, K.A.; Gavrilov, N.; Pašti, I.A.; Morávková, Z.; Acharya, U.; Unterweger, C.; Breitenbach, S.; Zhigunov, A.; Bober, P. Polyaniline-Metal Organic Framework (Fe-BTC) Composite for Electrochemical Applications. *Polymer* **2020**, *208*, 122945. [CrossRef]
49. Vujković, M.; Gavrilov, N.; Pašti, I.; Krstić, J.; Travas-Sejdic, J.; Ćirić-Marjanović, G.; Mentus, S. Superior Capacitive and Electrocatalytic Properties of Carbonized Nanostructured Polyaniline upon a Low-Temperature Hydrothermal Treatment. *Carbon* **2013**, *64*, 472–486. [CrossRef]
50. Gavrilov, N.; Pašti, I.A.; Vujković, M.; Travas-Sejdic, J.; Ćirić-Marjanović, G.; Mentus, S.V. High-Performance Charge Storage by N-Containing Nanostructured Carbon Derived from Polyaniline. *Carbon* **2012**, *50*, 3915–3927. [CrossRef]
51. Gavrilov, N.; Vujković, M.; Pašti, I.A.; Ćirić-Marjanović, G.; Mentus, S.V. Enhancement of Electrocatalytic Properties of Carbonized Polyaniline Nanoparticles upon a Hydrothermal Treatment in Alkaline Medium. *Electrochim. Acta* **2011**, *56*, 9197–9202. [CrossRef]
52. Breitenbach, S.; Unterweger, C.; Lumetzberger, A.; Duchoslav, J.; Stifter, D.; Hassel, A.W.; Fürst, C. Viscose-based Porous Carbon Fibers: Improving Yield and Porosity through Optimization of the Carbonization Process by Design of Experiment. *J. Porous Mater.* **2021**, *28*, 727–739. [CrossRef]
53. Breitenbach, S.; Lumetzberger, A.; Hobisch, M.A.; Unterweger, C.; Spirk, S.; Stifter, D.; Fürst, C.; Hassel, A.W. Supercapacitor Electrodes from Viscose-Based Activated Carbon Fibers: Significant Yield and Performance Improvement Using Diammonium Hydrogen Phosphate as Impregnating Agent. *C* **2020**, *6*, 17. [CrossRef]

C Journal of Carbon Research

MDPI

Article

High Energy Density Primary Lithium Battery with Fluorinated S-Doped Graphene

Marie Colin [1], Hani Farhat [1], Sam Chen [2], Elodie Petit [1], Emmanuel Flahaut [3], Katia Guérin [1] and Marc Dubois [1,*]

1 Institut de Chimie de Clermont-Ferrand (UMR 6296), CNRS, Université Clermont Auvergne, BP 10448, F-63178 Aubière, France; marie.colin@uca.fr (M.C.); hani.farhat@univ-pau.fr (H.F.); elodie.petit@uca.fr (E.P.); katia.araujo_da_silva@uca.fr (K.G.)
2 School of Environmental and Life Sciences, The University of Newcastle, Callaghan, NSW 2308, Australia; sam.chen@newcastle.edu.au
3 CIRIMAT, Université Toulouse 3 Paul Sabatier, Toulouse INP, CNRS, Université de Toulouse, 118 Route de Narbonne, CEDEX 09, F-31062 Toulouse, France; emmanuel.flahaut@univ-tlse3.fr
* Correspondence: marc.dubois@uca.fr

Abstract: Sulphur-doped graphene was fluorinated using molecular fluorine (F_2). First, the fluorination conditions were adapted in order to be mild enough to maintain S in the carbon lattice and form S-F bonds. An unusually weakened C-F bonding for an F/C ratio of 0.71 was then achieved, which allowed enhanced performances when used as a cathode in primary lithium batteries. The material prepared at a moderate fluorination temperature of 70 °C for a period of 60 min exhibits a high mid-discharge reduction potential of 3.11 V at 10 mA/g and a power density of 3605 W/kg at a discharge rate of 2C. These electrochemical properties make the fluorine/sulfur co-doped graphene a promising material.

Keywords: graphene; Co-doping; S-doping; fluorinated carbons; fluorination; primary lithium battery; cathode

1. Introduction

An expanded and/or nanometric structure can improve electrochemical performances, in particular by ensuring better ionic diffusion during the electrochemical reactions. In addition to a nanometric structure favorable to ionic diffusion, the electron density can be modulated by the presence of heteroelements, i.e., doping. For carbonaceous (nano)materials, the heteroelements most commonly used for doping are nitrogen and sulfur, and more rarely phosphorus and boron. In the present work, commercial graphene is used as the carbon matrix in order to combine both homogeneous doping and nanometric size. The presence of nitrogen atoms in graphene increases the conductivity of the material by creating active sites that will favor electron transfer [1]. These atoms can be linked to carbons in different ways: (i) forming a pyrrole type, an aromatic ring of five atoms, one of which is nitrogen; (ii) giving a pyridine type when nitrogen substitutes carbon in a benzene ring at the edge of pinholes or the edge of graphene sheets; or (iii) forming a graphitic type by substituting a carbon atom within the graphitic network [2,3]. Co-doping is possible when multiple types of heteroatoms are introduced into graphene, such as N, F or P, F co-doped graphene. It has been found that the doping of graphene with nitrogen or phosphorus leads to the weakening of C-F bonds [3,4]. N, S co-doping led to the surface defluorination of fluorinated carbons (CF_x) with weakened covalent C-F bonds [5]. To the best of our knowledge, studies on doping fluorinated graphene do not show evidence of bond formation between the N, P, or S heteroatom and fluorine [3,6].

Carbons doped with heteroatoms such as nitrogen, phosphorus, or boron are widely used for energy storage [4,5,7–11]. In particular, boron-doped fluorinated graphene (B-CF_x) delivers high capacities of up to 1200 mAh/g at 0.01C thanks to the defects created by the boron atoms in the graphene, which allow Li^+ ions to adsorb onto the defects, giving extra capacity [7]. B-CF_x then delivers a very high energy density of 2974 Wh/kg. Boron doping makes it possible to increase the fluorination rate while maintaining electrochemically active C-F bonds, preferably with weakened covalency, in order to guarantee a high reduction potential. Some research has been performed on sulfur doping. SF_6 has been used as a fluorinating agent for the preparation of fluorine and sulfur co-doped amorphous carbon films and fluorine and sulfur co-doped suspended sulfur graphene [12,13]; only covalent C-F bonds were formed, and sulfur was present in C-S-C groups, but no S-F bonds were formed. Peng et al. incorporated sulfur by manual grinding into fluorinated graphene, which formed C=S bonds [11]. In the present work, the direct gas/solid fluorination of sulfur-doped graphene was investigated with molecular fluorine gas (F2). It should be noted that fluorine/sulfur-doped materials have not been tested as cathodes in lithium batteries yet.

Fluorinated graphene exhibits electronic, thermal, and even tribological properties that differ from graphene, making it suitable for use in energy storage systems. The most common route of preparation for fluorinated graphene involves exfoliating fluorinated graphite (top-down approach). Bottom-up methods such as direct gaseous fluorination, plasma fluorination with CF_4 or SF_6, etc., are also used [14–17], as well as liquid-phase fluorination using solvents [18–20]. Fluorinated graphene can be obtained using graphite oxide as a precursor. The first work on fluorinating graphite oxide with molecular fluorine was carried out by Nakajima et al. [21]. The graphite oxide was also fluorinated with hydrofluoric acid at high temperatures [22]. Graphite oxide (GO) is composed of stacked sheets of graphene oxide (GO) and can be exfoliated to isolate those sheets. The shaping of graphene oxide films produces a light, flexible, and self-supporting material that can be used in energy storage systems. The presence of oxygen in CF_x can improve electrochemical performance; Mar et al. obtained higher potentials and capacities in the simultaneous presence of oxygen and fluorine, leading to increased mass-energy densities of 2398 Wh/kg compared with only 835 Wh/kg for a conventional CFx without oxygen [23]. We then seek to take advantage of the simultaneous presence of two heteroelements and sulfur in the present work.

After fluorinating the commercial sulfur-doped graphene with molecular fluorine gas sulfur, the product was characterized to study the morphology, structure, elemental distribution, and quantity of the heteroelements (sulfur, oxygen, and fluorine) within the material and the bonds formed after fluorination. Finally, they will be studied as a cathode electrode in a primary lithium battery in order to evidence the possible improvements thanks to the presence of both S and F.

2. Materials and Methods

2.1. Fluorination of Sulfur-Doped Graphene

Sulfur-doped graphene (Sigma Aldrich, Saint Louis, MO, USA) consists of 85–95% carbon, 2.0–4.0% sulfur, and less than 7.5% oxygen, according to the supplier's data. Sulfur-doped graphene (S-graphene) was fluorinated by static fluorination with molecular fluorine gas using the following procedure: fluorine gas was introduced into the reactor to obtain a relative pressure of 0.5 bar, and the reactor was then filled with nitrogen until atmospheric pressure was reached. An F_2/N_2 gas mixture (50/50 %vol.) was used for different fluorination temperatures and durations (Table 1). S-doped graphene exhibits both a low crystalline order and a high specific surface area (>500 m^2/g according to the product specification) that induce high reactivity towards molecular fluorine [24]; so, dilution of F_2 and the low fluorination temperature in the 25–140 °C range were used. The color of the compound changed from black to brown as the fluorination temperature increased, as is generally observed for CF_x [25].

Table 1. Fluorination temperatures and durations for sulfur-doped graphene.

Notation	Temperature (°C)	Duration (min)
F25-30 min	25	30
F70-30 min	70	30
F70-60 min	70	60
F140-60 min	140	60

2.2. Physico-Chemical Characterization

X-ray diffraction (XRD) patterns were recorded using a PANalytical X'PERT X-ray diffractometer with Cu(Kα) radiation. FTIR experiments were conducted with a Nicolet Summit (Thermoscientific, Waltham, MA, USA) spectrometer in an Attenuated Total Reflection (ATR) mode. For each spectrum, 128 scans with 4 cm^{-1} resolution were collected between 4000 and 450 cm^{-1}. ^{19}F and ^{13}C solid-state NMR experiments were performed using a 300 MHz Bruker Avance spectrometer at room temperature. A cross-polarization (CP)/magic-angle spinning (MAS) NMR probe operating with 2.5 mm and 4 mm rotors was used at 30 kHz and 10 kHz spinning rates for ^{19}F and ^{13}C measurements, respectively. A probe (Bruker, Billerica, MA, USA) with fluorine decoupling on a 4 mm rotor was used. For MAS spectra, a simple sequence was performed with a single $\pi/2$ pulse length of 4.0 and 3.5 µs for ^{19}F and ^{13}C, respectively. ^{13}C NMR was performed at a frequency of 73.4 MHz, and tetramethylsilane (TMS) was used as the reference. ^{19}F NMR was carried out with a frequency of 282.2 MHz and the spectra were externally referenced to CF_3COOH, and then to $CFCl^3$ ($\delta_{CF3COOH}$ = −78.5 ppm/$CFCl_3$). Transmission Electron Microscopy (TEM) samples were prepared with chloroform and ultrasonication. TEM analysis was performed using a JEOL JEM-F200 Multi-Purpose FEG-S/TEM operating at an accelerating voltage of 200 kV. X-ray photoelectron spectroscopy (XPS) spectra were recorded using an ESCALAB 250 Xi, Thermo Scientific, with a monochromated Al K alpha X-ray source (1486.68 eV) and a pass energy of 100 eV and a step energy of 1 eV for survey scans or 20 eV and a step energy of 0.1 eV for region scans. An electrochemical study was performed using both cyclic voltammetry and galvanostatic discharge. The positive electrode of the powdered fluorinated S-graphene compound was composed of the sample (80 wt.%, 2–3 mg), acetylene black (from Mersen, Courbevoie, France, 10 wt.%), to ensure the electronic conductivity and polyvinylidene difluoride (PVDF from Aldrich, 10 wt.%) as the binder. After stirring in propylene carbonate (PC), the mixture was spread uniformly onto a stainless-steel current-collector disk 12 mm in diameter. After PC evaporation, the disk was heated under vacuum at 40 °C and then 120 °C, for 1 h at each temperature, to remove traces of both water and solvent. The electrolyte was composed of lithium hexafluorophosphate ($LiPF_6$) salt dissolved in a mixture of propylene carbonate/ethylene carbonate/dimethyl carbonate (PC:EC:3DMC; 1:1:3 vol%). Three Celgard separators impregnated with the electrolyte were placed between the electrodes, and a lithium foil was used both as counter and reference electrodes. The button cells (CR2032) were assembled in an argon-filled dried glove box. Relaxation was performed for 5 h before any electrochemical process. Then, cyclic voltammetry was carried out at a scan rate of 0.01 mV.s^{-1} in the potential range between 1.5 V and 4.0 V at room temperature. Galvanostatic discharge curves were recorded at 10 mA/g until 1.5 V. Power tests were performed at different discharge rates considering 1C = 751 mA/g. The tests were carried out using an MPG apparatus from Biologic. For post-mortem ^{19}F NMR analysis, the cell was opened in an argon-filled dried glove box, and the electrode was rinsed with PC and dried before analysis.

3. Results

3.1. Physico-Chemical Characterization

X-ray diffractograms of S-graphene before and after fluorination are shown in Figure 1. Before fluorination, the S-graphene compound shows a diffraction peak at 25.9° for the (002) plane of the initial graphitic matrix. The planes diffract at an angle close to that of graphite (2θ~26.5°), which means that the S-graphene compound is not of a single-layer type but is rather multi-layer. The width of the peak and its lower intensity compared with graphite show that the material exhibits a weakly crystallized structure [26]. In other words, the compound has a lower degree of stacking order between the layers. The peak corresponding to the (002) plane is retained for the F25-60 min compound but has slightly shifted towards a smaller angle value, indicating a d-spacing of 0.36 nm for F25-60 min (compared to 0.35 nm for S-graphene). A new broad diffraction peak appeared near 14°; however, the fluorocarbon matrix is only slightly present in this material, demonstrating that fluorination is not very pronounced. From a fluorination temperature of 70 °C, the (002) peak of a stacked graphene phase disappears, evidencing the completion of fluorination.

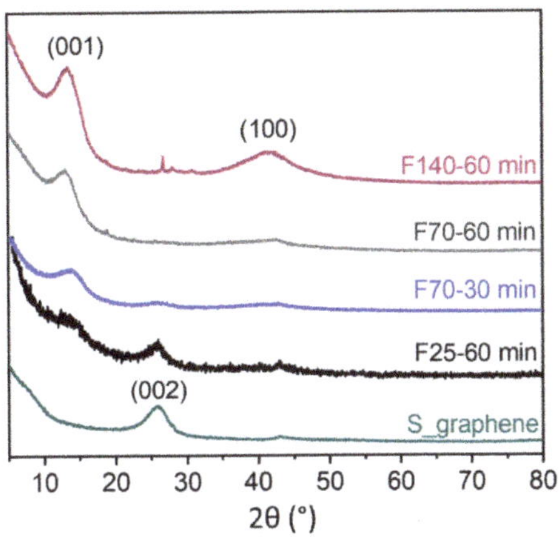

Figure 1. X-ray diffractograms of S-graphene compounds before and after fluorination.

Non-fluorinated and fluorinated compounds F70-60 min and F140-60 min were analyzed by transmission electron microscopy (Figure 2). TEM images of the compounds highlight a disordered structure, which becomes even more disorganized after fluorination at 70 °C and 140 °C. The TEM images are consistent with the XRD analysis.

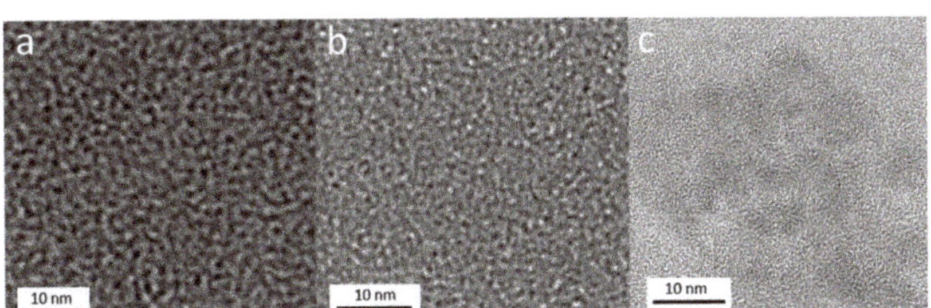

Figure 2. TEM images of non-fluorinated S-graphene compounds (**a**), F70-60 min (**b**), and F140-60 min (**c**).

Elemental maps of carbon, fluorine, oxygen, and sulfur have been acquired and are shown in Figure 3. The average atomic percentages of the elements extracted from the EDX measurements are summarized in Table 2 for S-graphene, F70-60 min, and F140-60 min. The elements C and F are clearly visible on the EDX maps for both fluorinated samples. The maps show that fluorine is homogeneously distributed within the graphene material. The chemical compositions show that oxygen in the S-graphene can still be detected after fluorination but in very small quantities. The analysis also shows the presence of sulfur only for the compound F70-60 min in contrast to F140-60 min. Quantification of the elements evidences a decrease in the carbon content from around 94% initially to 91% and 83% after fluorination at 70 °C and 140 °C, respectively, with fluorine appearing at 5% and 17%. The amount of sulfur decreases slightly after fluorination at 70 °C and completely after fluorination at 140 °C.

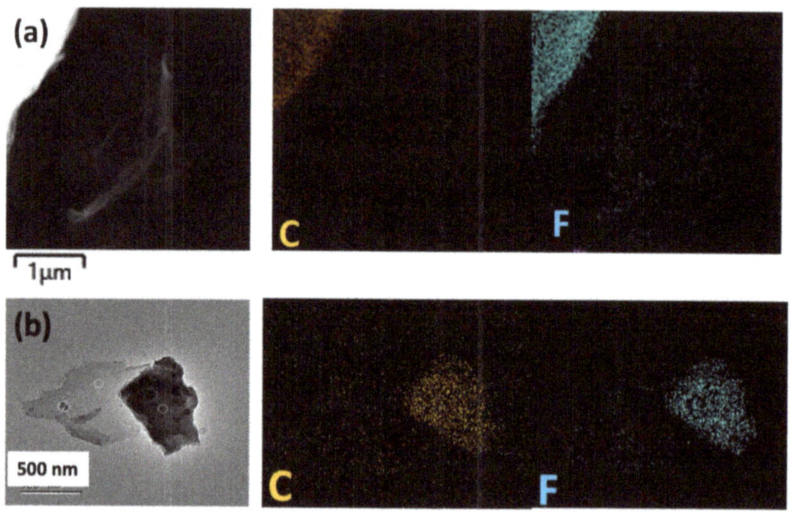

Figure 3. TEM images of compounds F70-60 min and elemental mapping of compounds of F70-60 min (**a**) and F140-60 min (**b**). S mappings are not shown because of the low S content.

Table 2. Chemical composition in C, O, S, and F of the compounds S-graphene, F70-60 min, and F140-60 min.

Compound	Chemical Composition			
	TEM-EDX			
	C	O	S	F
S-graphene	93.68	2.99	2.47	---
F70-60 min	90.75	1.93	2.00	5.02
F140-60 min	83.02	0.46	0.00	16.82

The infrared spectra of fluorinated S-graphene acquired with ATR mode are shown in Figure 4. We observe the appearance of a broad massif between 1200 and 1100 cm^{-1} after fluorination, which increases in intensity as the temperature and fluorination time increase. This cluster corresponds to the superimposition of bands related to the vibration mode of covalent C-F bonds at 1210 cm^{-1} and weakened covalent C---F bonds at 1160 cm^{-1}. At the highest temperature (140 °C), the material contains a majority of covalent C-F bonds, and the shoulder peak corresponding to the weakened covalent C---F bonds has weakened. CF_2 groups are also revealed by the vibrational band at 1320 cm^{-1}; their formation is favored by the edges of the graphene sheets. An unusual doublet band centered at 932 and 922 cm^{-1} was observed, with an absorption maximum for the F70-60 min compound. This doublet was identified in the case of compound SF4 and has been assigned to the vibration of S-F bonds [27]. Fluorination conditions that are too drastic, i.e., 140 °C for 1 h, result in the disappearance of the doublet assigned to S-F. Conversely, conditions that were too mild in terms of the duration (30 min at 70 °C) or temperature (1 h at 25 °C) were not sufficient to form a high amount of S-F bonds, as evidenced by the low intensity of the doublet on the IR spectrum.

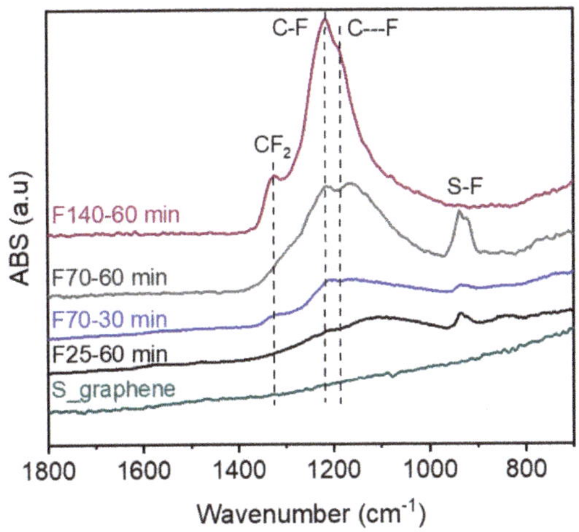

Figure 4. Infrared spectra of sulfur-doped graphene before and after fluorination under different conditions.

The ^{19}F and ^{13}C NMR spectra of fluorinated S-graphene are shown in Figure 5. In ^{19}F NMR, three main contributions are visible in the bulk corresponding to bonds involving fluorine atoms (Figure 6a). The C-F bonds and the CF_2 and CF_3 groups are highlighted by bands at −173, −120, and −78 ppm, respectively [25,28–30]. The covalency of the C-F bonds becomes stronger as the fluorination temperature increases, resulting in a chemical shift moving towards the lower values, i.e., from −150 to −180 ppm. Another band at 58 ppm is attributed to S-F bonds, already identified by infrared spectroscopy (Figure 5). The S-F bond band is more intense in the case of long fluorination (F25-60 min and F70-60 min) at a "moderate" temperature. The S-F bonds disappeared after fluorination at higher temperatures, i.e., 140 °C.

The ^{13}C NMR spectra (Figure 5b) do not show a band assigned to carbon-sulfur bonds. Within the NMR detection limit, our fluorinated materials therefore contain no C-S bonds (C-S-C groups) as reported in the literature [12,13]. Two main bands are present at 84 ppm for covalent C-F bonds and between 120 and 135 ppm for non-fluorinated C sp2 atoms. In a weakly fluorinated environment (F25-60 min), the sp^2 C atoms show a resonance at 122 ppm, close to that expected for graphite (120 ppm). When the amount of fluorine is higher (F140-60 min), this band moves towards higher chemical shifts, suggesting a decrease in the hyperconjugation; the more fluorinated the environment of a C-F bond, the less influence the neighboring C sp^2 has on the electron density around the ^{19}F nuclei. In other words, the C-F bond is strengthened [29,31]. In the case of a higher degree of fluorination, as for F140-60 min, the quantity of sp^2 C atoms decreases in favor of the formation of C-F bonds. The F/C fluorination rate of the compounds is calculated by deconvolution of the ^{13}C spectra ($F/C = (S_{C\text{-}F} + 2S_{CF2})/(S_{C\text{-}F} + S_{CF2} + S_{Csp2} + S_{Csp3})$ with S being the integrated surfaces of the bands). The F/C ratios are 0.22 ± 0.02 for F25-60 min, 0.51 ± 0.02 for F70-30 min, 0.72 ± 0.02 for F70-60 min, and 0.82 ± 0.02 for F140-60 min. The fluorination rate increases when the fluorination temperature is higher. It also increases for longer fluorination at the same temperature (70 °C). The compound fluorinated for 60 min has a higher F/C ratio than the compound fluorinated for 30 min. The $^{19}F \rightarrow ^{13}C$ cross-polarization (CP) NMR spectra of compounds F70-60 min and F140-60 min are shown in Figure 5c. This analysis is in good agreement with the ^{13}C and ^{19}F NMR analyses. CP favors the nuclear resonance of C directly bonded to fluorine atoms. After fluorination at 140 °C, the quantity of non-fluorinated sp^2 C is much lower, or even non-existent, in contrast to the compound fluorinated at 70 °C. CF2 groups are much more present after fluorination at 140 °C.

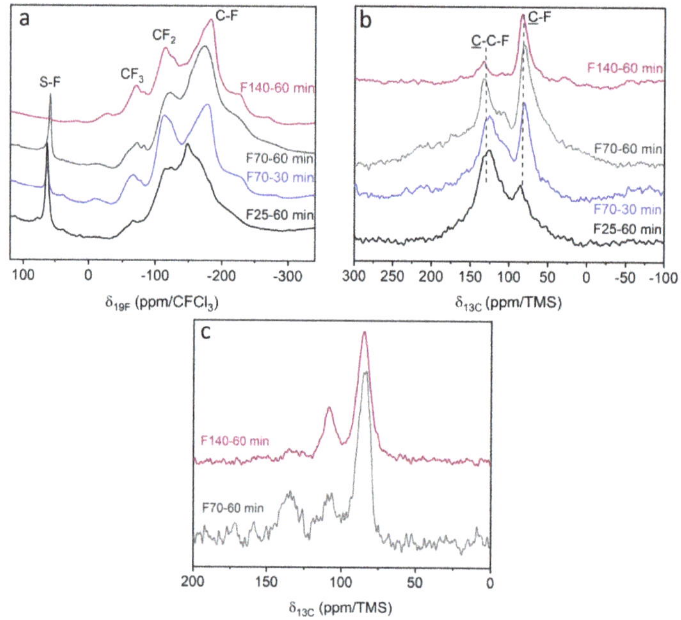

Figure 5. ^{19}F (**a**), ^{13}C (**b**), and $^{19}F \rightarrow ^{13}C$ cross-polarization (**c**) NMR spectra of sulfur-doped fluorinated graphene.

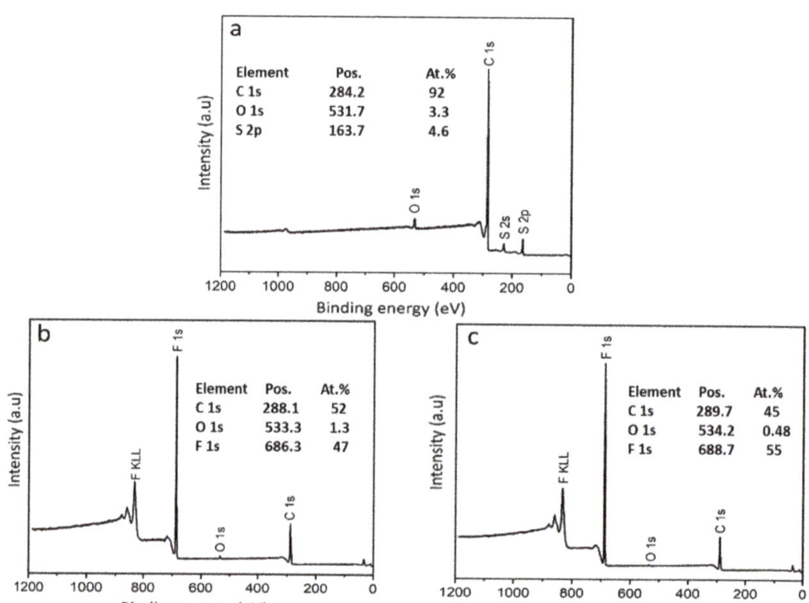

Figure 6. XPS spectra of S-graphene before (**a**) and after fluorination at 70 °C (**b**) and 140 °C (**c**) for 60 min.

The XPS survey spectra of the non-fluorinated and fluorinated compounds at 70 and 140 °C for 1 h are shown in Figure 6. The C1s, F1s, and S2p XPS spectra of the fluorinated compounds are given in Figure 7 and the experimental data from the analysis of the XPS spectra are given in Table 3. The as-received S-graphene contains 4.6 atomic (at.) % sulfur, indicated by the presence of the S2s and S2p peaks, which is in agreement with the supplier data. After fluorination at 70 and 140 °C, the C1s peak at 290.0 eV decreased significantly in intensity. The carbon content drops from 92 at.% for S-graphene to 52 at.% for F70-60 min and 45 at.% for F140-60 min. The introduction of fluorine is evident with a peak at 688.7 eV and 688.5 eV for F70-60 min and F140-60 min, respectively. The fluorine contents are 47 at.% for F70-60 min and 55 at.% for F140-60 min, which are over-expressed at the extreme surface of the material when the fluorination temperature is higher. The surface fluorination ratios F/C were 0.90 for F70-60 min and 1.21 for F140-60 min. The amount of oxygen, 3.3 at.% in the S-graphene, decreases after fluorination and when increasing the temperature. O is present at 1.30 at.% for F70-60 min and only at 0.48 at.% for F140-60 min. The C1s spectrum of F70-60 min (Figure 7a) can be deconvoluted into six contributions, three of which are in majorities. Most C-F bonds (62.0 at.%) show a binding energy of 290.0 eV, sp2 C atoms with a fluorinated environment (CF-C-CF) emerge at 287.5 eV, and CF_2 groups at 292.0 eV. The C1s spectrum of compound F140-60 min (Figure 7d) is also deconvoluted into six contributions, three of which are positioned at the same binding energies. C-F bonds are still dominant (63.9 at.%). The amount of sp^2 C (CF-C-CF) is lower after fluorination at 140 °C with 7.8 at.% compared with 18.9 at.% for F70-60 min. However, CF_2 groups are present in greater quantities for F140-60 min (24.0 at.% compared with 13.4 at.% for F70-60 min). The presence of fluorine is evident by an F1s peak located near 688 eV corresponding to covalent C-F bonds (Figure 7b,e) [32] for both compounds. Sulfur is present on the surface of the compound only for F70-60 min. The S2p spectra (Figure 7c) are deconvoluted into several contributions corresponding to two major groups, SOx and SFx. Based on several reports [33–35], the SOx environments range from 165.5 eV to 170.2 eV. In our case, we have three environments: the first environment is S 2p3/2 S1 (-S=O) at 165.7 eV associated with the S 2p1/2 orbital at 166.9 eV (hatched contribution) with an orbital splitting of 1.2 eV and an intensity ratio of 2. For the second environment corresponding to S 2p3/2 S2 (SO_3^{2-}) at 167.9 eV, the S 2p1/2 at 169.1 eV also exhibits the same orbital splitting and the same intensity ratio. For the third environment, S 2p3/2 S3 (SO_4^{2-}) at 169.4 eV 2p3/2 and 2p1/2 at 170.6 eV. As in the reference [36,37], the sum of the two orbitals (S2p3/2 and S2p1/2) of S 2p spectra are highlighted (Figure 7c and Table 3). The higher the oxygen coordination, the more the binding energy shifts towards higher energies.

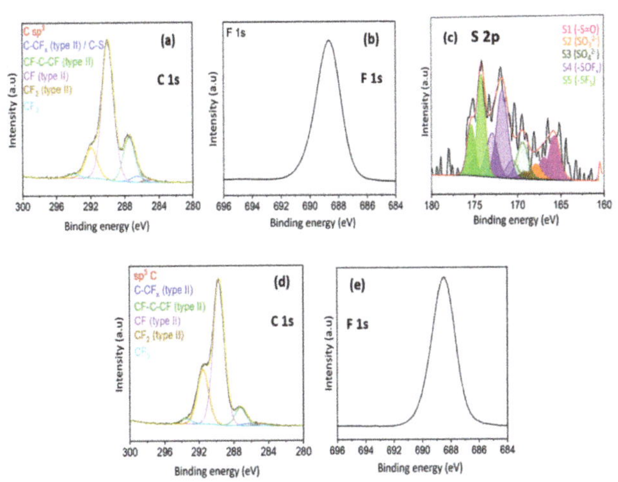

Figure 7. XPS C 1s and F 1s spectra of compounds F70-60 min (**a**,**b**) and F140-60 min (**d**,**e**) and S 2p spectrum of compound F70-60 min (**c**).

Table 3. Relative contents of carbon-, oxygen-, fluorine-, and sulfur-containing groups from the deconvolution of XPS spectra of compounds F70-60 min and F140-60 min. Because of the low sulfur content, no component of either C-S or S-F is considered.

Element	Assignment	F70-60 Min			F140-60 Min	
		Binding Energy (eV)		Relative Amount (at.%)	Binding Energy (eV)	Relative Amount (at.%)
C	C (sp^3)	285.0		0.62	284.7	0.43
	C̲ -CFx (type II) & C-S	286.6		1.4	286.3	0.45
	CF-C̲-CF (type II)	287.5		9.8	287.3	3.5
	C̲F (type II)	290.0		31.9	289.7	28.8
	CF_2 (type II)	292.0		6.9	291.6	10.8
	CF_3	294.0		0.91	293.5	1
O	O̲-C-F	535.1		1.57	534.96	0.45
	C-O̲	-		-	532.15	0.11
F	C-F	688.7		46.7	688.5	54.5
S	SO_x & SF_x	S 2p3/2	S 2p1/2		-	-
	S1 (-S=O)	165.7	166.9	0.02		
	S2 (SO_3^{2-})	167.9	169.1	0.01	-	-
	S3 (SO_4^{2-})	169.4	170.6	0.01	-	-
	S4 (-SOF_x)	171.7	172.9	0.04	-	-
	S5 (-SF_5)	174.2	175.4	0.03	-	-

For SFx environments [38], they fall in the higher energy range above 170.2 eV, depending on the degree of coordination with fluorine, with up to 178.5 eV for SF_6 as cited in the following article from the XPS database. In our case, we have three environments: the first environment is S 2p3/2 S4 (-SOF_X) at 171.7 eV associated with S 2p1/2 at 172.9 eV, with $x > 2$ but less than 5, an intermediate environment between (S 2p3/2 (-SOF_2) at 170.2 eV) and SF_5. The second environment is 2p3/2 S5 (SF_5) at 174.2 eV and S 2p1/2 at 175.4 eV.

3.2. Electrochemical Tests

In order to study the effect of sulfur and the unusual S-F bond on the reversibility of Li/CF_x batteries, the compound fluorinated at 70 °C for 1 h (F70-60 min) was tested as a positive electrode in primary lithium batteries using lithium as the anode. The choice fell to the material with the highest fluorination rate ($F/C = 0.72 \pm 0.02$) but that retains the S-F bonds formed during fluorination.

Cyclic voltammetry of compound F70-60 min recorded at 0.01 mV/s with $LiPF_6$ in the EC/PC/3DMC 1M electrolyte is shown in Figure 8. This electrochemical analysis technique highlights the different types of bonds determined during previous analyses by both infrared spectroscopy and solid-state NMR.

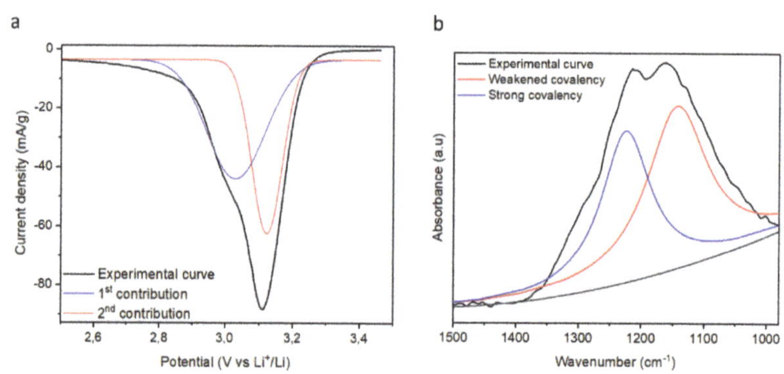

Figure 8. (**a**) Voltammogram of compound F70-60 min recorded at 0.01 mV/s with 1M $LiPF_6$ in EC/PC/3DMC electrolyte, (**b**) IR spectrum of F70-60 min.

The two components centered at 3.12 and 3.04 V are necessary to fit the reduction branch (Figure 8a). Each is related to one type of C-F bond: the stronger the covalence, the lower the discharge potential. Those data are in perfect agreement with the two IR vibration bands, which are assigned to pure and weakened covalent C-F bonds.

Galvanostatic discharge makes it possible to quantitatively evaluate the electrochemical properties of the material in battery applications via the delivered capacity. The galvanostatic discharge curve at 10 mA/g up to 1.5 V for the F70-60 min compound is shown in Figure 9. Compound F70-60 min has a theoretical capacity of 751 mAh/g considering the F/C fluorination rate of 0.72 calculated from 13C NMR. The experimental discharge curve shows a flat profile with a single plateau, whereas two steps were expected at distinct potentials corresponding to the breaking of the two types of bond, as can be seen by cyclic voltammetry (Figure 9). The galvanostatic discharge, applied under these conditions, does not allow the different C-F bonds to be discriminated. The material delivers a capacity of 773 mAh/g through the reduction of up to 1.5 V, which is close to the theoretical value considering the experimental errors. A reduction in the C-F bonds occurred at a half discharge potential of 3.10 V. This value is high for a material with covalent C-F bonds and is more similar to the reduction potential of bonds with weakened covalency [39]. The high potential may also be due to the sp^2 hybridized carbon atoms, which ensure better electronic conduction and thus contribute to increasing the potential. A high potential is observed for the case of fluorinated graphite prepared at room temperature with a catalytic gaseous atmosphere [40]. For comparison, the material in question is composed of weakened C-F bonds (C---F) and has an F/C fluorination rate of approximately 0.5, which is lower than for F70-60 min. The level of non-fluorinated sp^2 carbons and the resulting hyperconjugation do not explain the high potential of F70-60 min; this point will be discussed later.

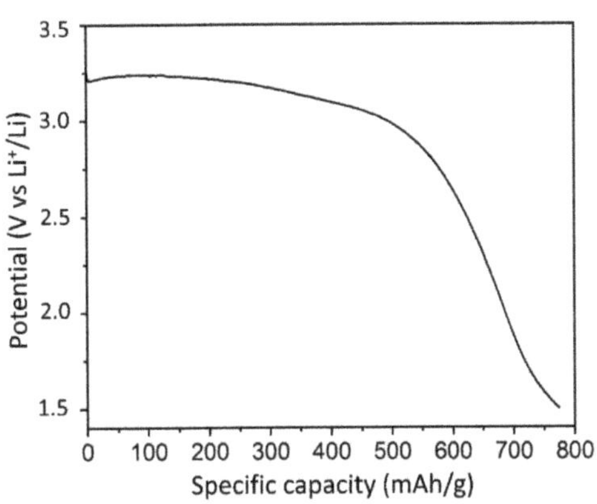

Figure 9. Galvanostatic discharge curve for F70-60 min at 10 mA/g with 1M $LiPF_6$ in EC/PC/3DMC.

The reduction potential of the F70-60 min compound is advantageously high since a high potential coupled with a high capacity will deliver a high energy density. The galvanostatic discharge curves for the F70-60 min compound at different current densities are shown in Figure 10. The applied current densities and experimental electrochemical data such as the capacity (C_{exp}), half-discharge potential ($E_{1/2}$), specific energy, power density, and faradic efficiency are summarized in Table 4.

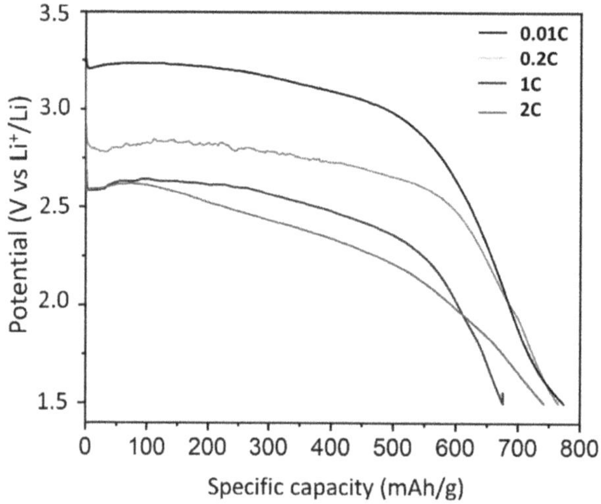

Figure 10. Galvanostatic discharge curves for F70-60 min at different current densities.

Table 4. Electrochemical data for F70-60 min obtained at different reduction current densities.

C Rate	Current Densities (mA/g)	$C_{exp} \pm 10\%$ (mAh/g)	$E_{1/2}$ (V)	Specific Energy (Wh/kg)	Power Density (W/kg)	Faradic Yield (%) ± 10%
0.01C	10	773	3.10	2396	31	103
0.2C	150	764	2.74	2093	411	102
1C	751	675	2.54	1715	1908	90
2C	1502	741	2.40	1778	3605	99

The discharge curves show a conventional profile for CF_x as the discharge regime increases with a single plateau corresponding to the breaking of the C-F bonds. More unusually, the material delivers similar capacities whatever the applied current density up to 2C, with faradic efficiencies close to 100%. The potential remains fairly high, i.e., 2.40 V at 2C, enabling high specific energy densities of 1778 Wh/kg to be obtained. In addition, in contrast to conventional CF_x, the potential drop at the beginning of the discharge is non-existent thanks to the sp^2 C present in the material, which ensures better electronic conduction.

The reduction potential of B-CF_x is between 2.75 and 2.90 V at 0.01C with the 1M $LiPF_6$ dissolved in the EC/DMC electrolyte, depending on the fluorination degree of the material [7]. These potentials are still lower than the potential of 3.11 V obtained for the compound F70-60 min at 0.01C. However, it is difficult to compare electrochemical systems directly if they do not involve the same electrolyte and electrode formulation. CF_x doped with nitrogen and sulfur (N,S-CF_x) on the surface achieved high electrochemical performance up to discharge rates equivalent to 70C [5]. However, the reduction potential is lower in the case of N,S-CF_x. For example, at a discharge rate of 0.1C, N,S-CF_x delivers a capacity of 787 mAh/g at a potential of 2.59 V, resulting in an energy density of 2038 Wh/kg. In our case, the F70-60 min compound, reduced to a higher discharge rate of 0.2C, delivers a capacity of 764 mAh/g at a potential of 2.74 V, giving an energy density of 2093 Wh/kg. In order to compare the performance of the two materials at the same discharge rate, we will consider the equivalent discharge rate at 1C. The precise values of the capacities and potentials obtained at 1C for N,S-CF_x are not given by Zhu et al. but the reduction potential is estimated to be around 2.40 V, whereas our F70-60 min material reduces at a potential of 2.54 V at 1C. In the case of CF_x doped with phosphorus (P-CF_x), the material demonstrates capacities delivered up to 20C [4]. Polarization of the material at the start of the discharge is present from the 2C regime onwards, unlike the F70-60 min material, which does not show any. Electronic conduction therefore seems to be better in our case. In addition, the average reduction potential is also lower for P-CF_x, with a potential of 2.5 V at 0.1C giving an energy density of 2025 Wh/kg. At 2C, the potential appears to be close to 2.3 V, 0.1 V lower than F70-60 min. However, it is important to take into account the differences in the electrode formulation and electrolyte used, which can affect the electrochemical results. The electrode formulation using P-CF_x or N,S-CF_x involves vertically grown carbon fibers of the dimension 1D, which should provide better electronic conduction than the 0D spherical carbon used in our case (carbon black). In addition, the 1M LiBF4 dissolved in a PC/DMC mixture electrolyte should also lead to better electrochemical performance thanks to the salt used, which proved to be better than the $LiPF_6$ we used. Despite these two points, our material exhibits better performance and a better reduction potential when the most "favorable" conditions are not used. The F70-60 min compound seems to demonstrate better electrochemical performance in terms of the reduction potential, which leads to higher energy densities. The F70-60 min compound also delivers an interesting power density of 3605 W/kg at a discharge rate of 2C.

Cyclic voltammetry and galvanostatic discharge showed a high half-discharge potential, while the types of C-F bonds determined during physicochemical characterization tended to show high covalency for C-F bonds. In the literature, graphs representing the half-discharge reduction potential as a function of the ^{19}F chemical shift of C-F bonds have been drawn up. The figure highlights the correlation between these two parameters: the more negative the chemical shift of the C-F bonds, the stronger the covalency of the C-F bonds, and the lower the reduction potential (Figure 11, [40]). If the data for compound F70-60 min are placed on this type of graph, represented by a star, it can be seen that they lie significantly outside the established trend curves. (Figure 11) The reduction potential of the material appears to be much higher than expected in relation to its chemical shift in ^{19}F NMR, i.e., according to the covalency of the C-F bonds of which it is composed. An effect of the S-F bonds and/or doping on the battery's discharge potential can thus be indirectly demonstrated. Once again, materials with identical electrode and electrolyte formulations and equal current regimes should be compared.

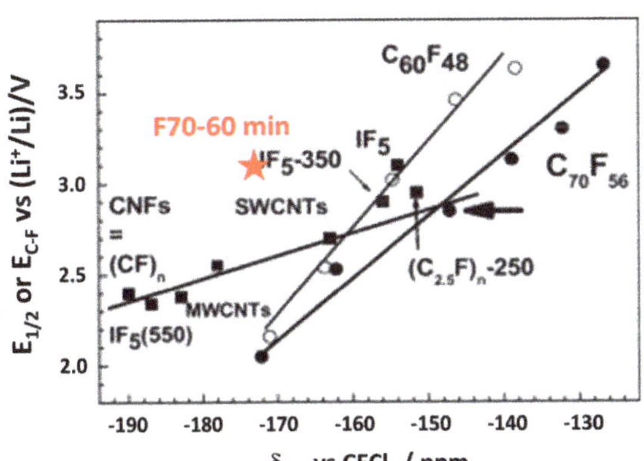

Figure 11. Reduction potential at mid-discharge as a function of the ^{19}F chemical shift of the C-F bonds of several materials taken from [40]. The cells used 1M $LiClO_4$ in PC electrolyte. The compound F70-60 min is represented by the red star. IF_5 means room temperature graphite prepared with IF_5 catalyst and its post-treatment temperature is added, e.g., IF_5-350 was post-fluorinated with pure F_2 gas at 350 °C. The same notation is used for graphite fluoride with $(C_{2.5}F)_n$ structural type.

An ^{19}F NMR post-mortem analysis was carried out on the electrode, i.e., the mixture of F70-60 min, PVDF, and acetylene black, after its preparation without contact with the electrolyte (Figure 12). The spectrum of the electrode showed several peaks. Firstly, there is a doublet at around −93 ppm corresponding to the CF_2 groups in the PVDF used for the electrode formulation. Two other changes can be noted:

- The position of the peak corresponding to the C-F bonds changes to a lower chemical shift. Initially, the C-F bonds resonate at −175 ppm in powdered F70-60 min, indicating the presence of covalent C-F bonds. After electrode formulation, the chemical shift of the C-F bonds fell sharply to −155 ppm. The electron conduction provided by the black acetylene in the electrode interfered with the NMR measurements, even on the ^{19}F nuclei. The peak of CF_2 in PVDF also shifted. In addition, the mass of active material is lower in the rotor (close to 2 mg, whereas around 10 mg are compacted in the rotor for powdered F70-60 min). This results in a lower signal-to-noise ratio. It is therefore difficult to say whether the strength of the C-F bond in the electrode material has been weakened during the electrode formulation.
- If the ratios of the intensities of the peaks linked to the S-F bonds (at 52 ppm) and C-F (−155 or −175 ppm) are considered for the spectra of the electrode and the initial powder, a decrease is noted after the formulation. Some of the S-F bonds were broken during electrode formulation. The unassigned peak at 10 ppm would be related to a decomposition product.

The ^{19}F NMR spectrum of the electrode after its discharge to 1.5 V was recorded (Figure 12). No peak is visible between −150 and −190 ppm, which means that all the C-F bonds were broken during the reduction. The S-F bonds were also completely reduced as evidenced by the absence of the peak at 52 ppm. The electrochemical reduction in F70-60 min formed LiF, which is characterized by an intense peak at −207 ppm.

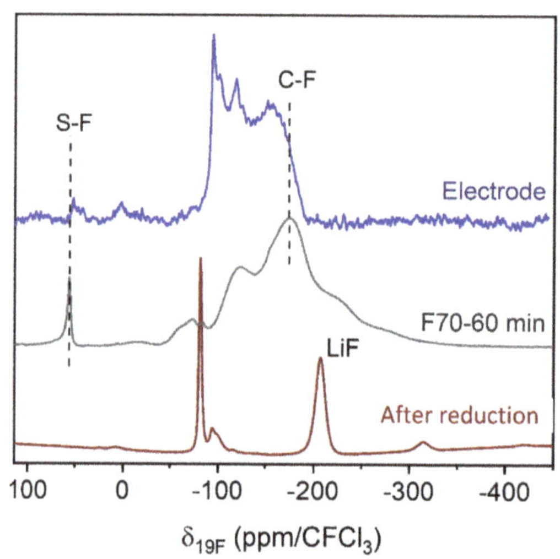

Figure 12. ^{19}F NMR spectra of F70-60 min (grey), of the electrode before any electrochemical process (blue), and of the electrode after electrochemical reduction of up to 1.5 V (red).

4. Conclusions

Sulfur doping increases the electrochemical performance of fluorinated graphene in primary lithium batteries by weakening the C-F bonds while allowing a high enough fluorination rate to deliver high capacities, unlike CF_x composed of semi-ionic C---F bonds, which have a lower fluorination rate and therefore lower delivered capacities [33,40]. Fluorination led to the formation of S-F bonds, which have never been reported in the literature. The formation of these bonds is favored at a moderate fluorination temperature of 70 °C for a period of 60 min but can be eliminated by stronger fluorination conditions. The F70-60 min compound resulting from these mild fluorination conditions has an F/C ratio of 0.71 and covalent C-F bonds but exhibits a high mid-discharge reduction potential of 3.11 V at 10 mA/g. The origin of this abnormally high potential is still ambiguous, and the role of S-F bonds and sulfur doping has yet to be clarified. Moreover, a power density of 3605 W/kg was obtained at a discharge rate of 2C. These electrochemical properties make the F,S co-doped graphene a promising material, and improvements concerning the formulation and chemical composition (e.g., changes in the S content) would confirm the high potentialities. Moreover, doping can be extended to other heteroelements such as B or P in order to tailor the properties, opening a new class of cathode materials for primary lithium batteries.

Author Contributions: Conceptualization, M.D., M.C., S.C. and K.G.; methodology, M.D. and K.G.; formal analysis, M.D., M.C., S.C., H.F., E.F., E.P. and K.G.; investigation, data curation, M.D., M.C., S.C., H.F., E.F., E.P. and K.G.; writing—original draft preparation, M.D., M.C., S.C., H.F., E.F. and K.G.; writing—review and editing, M.D., M.C., S.C., H.F., E.F. and K.G.; visualization, supervision, M.D. and K.G.; project administration, M.D.; funding acquisition, M.D. All authors have read and agreed to the published version of the manuscript.

Funding: This research received no external funding.

Data Availability Statement: Data is contained within the article.

Acknowledgments: M.D.: K.G. and M.C. acknowledge I-Site CAP2025, Université Clermont Auvergne, and Clermont Auvergne Métropole for support via the Académie CAP20-25; S.C. acknowledges the funding from the Australian Research Council (DP230100542), AMP Foundation's Tomorrow Fund, and the support from the University of Newcastle. E.F and M.D. acknowledge financial support from CNRS IRP C-FUNCTION.

Conflicts of Interest: The authors declare no conflicts of interest.

References

1. Panja, T.; Bhattacharjya, D.; Yu, J.-S. Nitrogen and phosphorus co-doped cubic ordered mesoporous carbon as a supercapacitor electrode material with extraordinary cyclic stability. *J. Mater. Chem. A* **2015**, *3*, 18001–18009. [CrossRef]
2. Zhuang, S.; Lee, E.S.; Lei, L.; Nunna, B.B.; Kuang, L.; Zhang, W. Synthesis of nitrogen-doped graphene catalyst by high-energy wet ball milling for electrochemical systems: Synthesis of N-G catalyst by high-energy wet ball milling. *Int. J. Energy Res.* **2016**, *40*, 2136–2149. [CrossRef]
3. Zhang, G.; Colin, M.; Yang, X.; Sun, S.; Dodelet, J.-P.; Dubois, M. C-F bonding in fluorinated N-Doped carbons. *Appl. Surf. Sci.* **2022**, *577*, 151721. [CrossRef]
4. Zhu, D.; Yuan, J.; Dai, Y.; Peng, Y.; Li, W.; Zhang, F.; Li, A.; Zhang, J. High-Rate Performance of Fluorinated Carbon Material Doped by Phosphorus Species for Lithium-Fluorinated Carbon Battery. *Energy Technol.* **2022**, *10*, 2200155. [CrossRef]
5. Zhu, D.; Yuan, J.; Wang, T.; Dai, Y.; Peng, Y.; Li, W.; Li, A.; Zhang, J. A novel one-step method to prepare N, S Co-doped sub-fluorinated carbon electrode materials for ultrahigh-rate lithium-fluorinated carbon battery. *J. Power Sources* **2022**, *551*, 232188. [CrossRef]
6. Wang, X.; Sun, G.; Routh, P.; Kim, D.-H.; Huang, W.; Chen, P. Heteroatom-doped graphene materials: Syntheses, properties and applications. *Chem. Soc. Rev.* **2014**, *43*, 7067–7098. [CrossRef] [PubMed]
7. Wang, K.; Feng, Y.; Kong, L.; Peng, C.; Hu, Y.; Li, W.; Li, Y.; Feng, W. The Fluorination of Boron-Doped Graphene for CF_x Cathode with Ultrahigh Energy Density. *Energy Environ. Mater.* **2022**, *6*, e12437. [CrossRef]
8. Shen, W.; Wang, C.; Xu, Q.; Liu, H.; Wang, Y. Nitrogen-Doping-Induced Defects of a Carbon Coating Layer Facilitate Na-Storage in Electrode Materials. *Adv. Energy Mater.* **2015**, *5*, 1400982. [CrossRef]
9. Li, Y.; Chen, M.; Liu, B.; Zhang, Y.; Liang, X.; Xia, X. Heteroatom Doping: An Effective Way to Boost Sodium Ion Storage. *Adv. Energy Mater.* **2020**, *10*, 2000927. [CrossRef]
10. Wan, H.; Shen, X.; Jiang, H.; Zhang, C.; Jiang, K.; Chen, T.; Shi, L.; Dong, L.; He, C.; Xu, Y.; et al. Biomass-derived N/S dual-doped porous hard-carbon as high-capacity anodes for lithium/sodium ions batteries. *Energy* **2021**, *231*, 121102. [CrossRef]
11. Peng, S.; Yan, S.; Wang, N.; Nan, W.; Wang, J.; Chen, X.; Wang, C.; Qi, X.; Dai, S. Fluorinated graphene/sulfur hybrid cathode for high energy and high power density lithium primary batteries. *RSC Adv.* **2018**, *8*, 12701–12707. [CrossRef] [PubMed]
12. Wang, F.; Wang, L.; Xue, Q. Fluorine and sulfur co-doped amorphous carbon films to achieve ultra-low friction under high vacuum. *Carbon* **2016**, *96*, 411–420. [CrossRef]
13. Struzzi, C.; Sezen, H.; Amati, M.; Gregoratti, L.; Reckinger, N.; Colomer, J.-F.; Snyders, R.; Bittencourt, C.; Scardamaglia, M. Fluorine and sulfur simultaneously co-doped suspended graphene. *Appl. Surf. Sci.* **2017**, *422*, 104–110. [CrossRef]
14. Plšek, J.; Drogowska, K.A.; Fridrichová, M.; Vejpravová, J.; Kalbáč, M. Laser-ablation-assisted SF6 decomposition for extensive and controlled fluorination of graphene. *Carbon* **2019**, *145*, 419–425. [CrossRef]
15. Yu, X.; Lin, K.; Qiu, K.; Cai, H.; Li, X.; Liu, J.; Pan, N.; Fu, S.; Luo, Y.; Wang, X. Increased chemical enhancement of Raman spectra for molecules adsorbed on fluorinated reduced graphene oxide. *Carbon* **2012**, *50*, 4512–4517. [CrossRef]

16. Wang, B.; Wang, J.; Zhu, J. Fluorination of Graphene: A Spectroscopic and Microscopic Study. *ACS Nano* **2014**, *8*, 1862–1870. [CrossRef] [PubMed]
17. Tahara, K.; Iwasaki, T.; Matsutani, A.; Hatano, M. Effect of radical fluorination on mono- and bi-layer graphene in Ar/F_2 plasma. *Appl. Phys. Lett.* **2012**, *101*, 163105. [CrossRef]
18. Gao, X.; Tang, X. Effective reduction of graphene oxide thin films by a fluorinating agent: Diethylaminosulfur trifluoride. *Carbon* **2014**, *76*, 133–140. [CrossRef]
19. Samanta, K.; Some, S.; Kim, Y.; Yoon, Y.; Min, M.; Lee, S.M.; Park, Y.; Lee, H. Highly hydrophilic and insulating fluorinated reduced graphene oxide. *Chem. Commun.* **2013**, *49*, 8991. [CrossRef]
20. Jahanshahi, M.; Kowsari, E.; Haddadi-Asl, V.; Khoobi, M.; Bazri, B.; Aryafard, M.; Lee, J.H.; Kadumudi, F.B.; Talebian, S.; Kamaly, N.; et al. An innovative and eco-friendly modality for synthesis of highly fluorinated graphene by an acidic ionic liquid: Making of an efficacious vehicle for anti-cancer drug delivery. *Appl. Surf. Sci.* **2020**, *515*, 146071. [CrossRef]
21. Nakajima, T.; Mabuchi, A.; Hagiwara, R. A new structure model of graphite oxide. *Carbon* **1988**, *26*, 357–361. [CrossRef]
22. Jankovský, O.; Šimek, P.; Sedmidubský, D.; Matějková, S.; Janoušek, Z.; Šembera, F.; Pumera, M.; Sofer, Z. Water-soluble highly fluorinated graphite oxide. *RSC Adv.* **2014**, *4*, 1378–1387. [CrossRef]
23. Mar, M.; Dubois, M.; Guérin, K.; Batisse, N.; Simon, B.; Bernard, P. Tuning fluorine and oxygen distribution in graphite oxifluorides for enhanced performances in primary lithium battery. *Carbon* **2019**, *141*, 6–15. [CrossRef]
24. Chatenet, M.; Berthon-Fabry, S.; Ahmad, Y.; Guérin, K.; Colin, M.; Farhat, H.; Frezet, L.; Zhang, G.; Dubois, M. Fluorination and its Effects on Electrocatalysts for Low-Temperature Fuel Cells. *Adv. Energy Mater.* **2023**, *13*, 2204304. [CrossRef]
25. Touhara, H.; Okino, F. Property control of carbon materials by fluorination. *Carbon* **2000**, *38*, 241–267. [CrossRef]
26. Siburian, R.; Sihotang, H.; Raja, S.L.; Supeno, M.; Simanjuntak, C. New Route to Synthesize of Graphene Nano Sheets. *Orient. J. Chem.* **2018**, *34*, 182–187. [CrossRef]
27. Yamamoto, H.; Matsumoto, K.; Matsuo, Y.; Sato, Y.; Hagiwara, R. Deoxofluorination of graphite oxide with sulfur tetrafluoride. *Dalton Trans.* **2020**, *49*, 47–56. [CrossRef]
28. Dubois, M.; Giraudet, J.; Guérin, K.; Hamwi, A.; Fawal, Z.; Pirotte, P.; Masin, F. EPR and Solid-State NMR Studies of Poly(dicarbon monofluoride) $(C_2F)_n$. *J. Phys. Chem. B* **2006**, *110*, 11800–11808. [CrossRef]
29. Giraudet, J.; Dubois, M.; Guérin, K.; Delabarre, C.; Hamwi, A.; Masin, F. Solid-State NMR Study of the Post-Fluorination of $(C_{2.5}F)_n$ Fluorine−GIC. *J. Phys. Chem. B* **2007**, *111*, 14143–14151. [CrossRef]
30. Zhang, W.; Dubois, M.; Guérin, K.; Bonnet, P.; Kharbache, H.; Masin, F.; Kharitonov, A.P.; Hamwi, A. Effect of curvature on C–F bonding in fluorinated carbons: From fullerene and derivatives to graphite. *Phys. Chem. Chem. Phys.* **2010**, *12*, 1388–1398. [CrossRef]
31. Sato, Y.; Itoh, K.; Hagiwara, R.; Fukunaga, T.; Ito, Y. Short-range structures of poly(dicarbon monofluoride) (C2F)n and poly(carbon monofluoride) (CF)n. *Carbon* **2004**, *42*, 2897–2903. [CrossRef]
32. Cheng, L.; Jandhyala, S.; Mordi, G.; Lucero, A.T.; Huang, J.; Azcatl, A.; Addou, R.; Wallace, R.M.; Colombo, L.; Kim, J. Partially Fluorinated Graphene: Structural and Electrical Characterization. *ACS Appl. Mater. Interfaces* **2016**, *8*, 5002–5008. [CrossRef] [PubMed]
33. Dubois, M.; Guérin, K.; Zhang, W.; Ahmad, Y.; Hamwi, A.; Fawal, Z.; Kharbache, H.; Masin, F. Tuning the discharge potential of fluorinated carbon used as electrode in primary lithium battery. *Electrochim. Acta* **2012**, *59*, 485–491. [CrossRef]
34. Feng, X.; Xu, M.; Guo, N.; Ma, R.; Yan, L.; Cai, L.; Jia, D.; Ai, L.; Wang, L. Dual-Salt-Induced Hierarchical Porous Structure in a Carbon Sheet for High Performance Supercapacitors. *Langmuir* **2023**, *39*, 6865–6873. [CrossRef] [PubMed]
35. Li, M.; Liu, C.; Zhao, H.; An, H.; Cao, H.; Zhang, Y.; Fan, Z. Tuning sulfur doping in graphene for highly sensitive dopamine biosensors. *Carbon* **2015**, *86*, 197–206. [CrossRef]
36. Kiciński, W.; Dziura, A. Heteroatom-doped carbon gels from phenols and heterocyclic aldehydes: Sulfur-doped carbon xerogels. *Carbon* **2014**, *75*, 56–67. [CrossRef]
37. Lacey, M.J.; Yalamanchili, A.; Maibach, J.; Tengstedt, C.; Edström, K.; Brandell, D. The Li-S battery: An investigation of redox shuttle and self-discharge behaviour with $LiNO_3$-containing electrolytes. *RSC Adv.* **2015**, *6*, 3632–3641. [CrossRef]
38. Barlow, A.J.; Quinton, J.S. Radio frequency SF6 plasma modified single-walled carbon nanotubes: Synchrotron spectroscopy and plasma characterisation studies. In Proceedings of the IEEE 2010 International Conference on Nanoscience and Nanotechnology (ICONN), Sydney, Australia, 22–26 February 2010; pp. 120–123. [CrossRef]
39. Lindberg, B.J.; Hamrin, K.; Johansson, G.; Gelius, U.; Fahlman, A.; Nordling, C.; Siegbahn, K. Molecular Spectroscopy by Means of ESCA II. Sulfur compounds. Correlation of electron binding energy with structure. *Phys. Scr.* **1970**, *1*, 286–298. [CrossRef]
40. Delabarre, C.; Guérin, K.; Dubois, M.; Giraudet, J.; Fawal, Z.; Hamwi, A. Highly fluorinated graphite prepared from graphite fluoride formed using BF3 catalyst. *J. Fluor. Chem.* **2005**, *126*, 1078–1087. [CrossRef]

C Journal of Carbon Research

Article

Plasma-Treated Cobalt-Doped Nanoporous Graphene for Advanced Electrochemical Applications

Florian Knabl [1,*], Nikolaos Kostoglou [1,*], Ram K. Gupta [2], Afshin Tarat [3], Steven Hinder [4], Mark Baker [4], Claus Rebholz [1,5] and Christian Mitterer [1]

1 Department of Materials Science, Montanuniversität Leoben, 8700 Leoben, Austria; claus@ucy.ac.cy (C.R.); christian.mitterer@unileoben.ac.at (C.M.)
2 Department of Chemistry, Pittsburg State University, Pittsburg, KS 66762, USA; rgupta@pittstate.edu
3 Loginns GmbH, 9020 Klagenfurt, Austria
4 Department of Mechanical Engineering Sciences, University of Surrey, Guildford GU2 7XH, UK; s.hinder@surrey.ac.uk (S.H.); m.baker@surrey.ac.uk (M.B.)
5 Department of Mechanical and Manufacturing Engineering, University of Cyprus, Nicosia 1678, Cyprus
* Correspondence: florian.knabl@unileoben.ac.at (F.K.); nikolaos.kostoglou@unileoben.ac.at (N.K.)

Abstract: Metal–carbon nanocomposites are identified as key contenders for enhancing water splitting through the oxygen evolution reaction and boosting supercapacitor energy storage capacitances. This study utilizes plasma treatment to transform natural graphite into nanoporous few-layer graphene, followed by additional milling and plasma steps to synthesize a cobalt–graphene nanocomposite. Comprehensive structural characterization was conducted using scanning and transmission electron microscopy, X-ray diffraction, Raman spectroscopy, gas sorption analysis and X-ray photoelectron spectroscopy. Electrochemical evaluations further assessed the materials' oxygen evolution reaction and supercapacitor performance. Although the specific surface area of the nanoporous carbon decreases from 780 to 480 m^2/g in the transition to the resulting nanocomposite, it maintains its nanoporous structure and delivers a competitive electrochemical performance, as evidenced by an overpotential of 290 mV and a Tafel slope of 110 mV/dec. This demonstrates the efficacy of plasma treatment in the surface functionalization of carbon-based materials, highlighting its potential for large-scale chemical-free application due to its environmental friendliness and scalability, paving the way toward future applications.

Keywords: graphene; plasma; cobalt; nanocomposites; nanoporous powders; water splitting; oxygen evolution reaction; supercapacitor

1. Introduction

The ever-increasing global demand for energy, coupled with the climate crisis from emissions of carbon dioxide and other greenhouse gases, underscores the urgent need for solutions to ensure a more sustainable future. Significant efforts are underway to develop materials that facilitate efficient and environmentally friendly energy conversion and storage. Hydrogen, as a clean energy carrier synthesized ideally through water electrolysis, holds promise for a sustainable future. Electrolytic water splitting, encompassing both the hydrogen evolution reaction (HER) and the oxygen evolution reaction OER) [1], requires an effective catalyst to enhance kinetics and cost-effectiveness. Meanwhile, supercapacitors are considered to be another important cornerstone of advanced clean energy technologies. They offer substantial power densities for energy storage and rapid charging and discharging, utilizing the formation of electric double layers and electrochemical pseudocapacitance [2].

Carbon-based nanoporous materials are prime candidates for both OER [3] and supercapacitor applications [4,5] due to their superior properties, such as high chemical and mechanical stability, excellent electrical conductivity and expansive specific surface areas with adjustable pore size distributions. Enhancing these intrinsic properties through surface functionalization is a key area of current research [6], offering the potential for even more optimized performance in energy-related applications. Cobalt nanoparticles [7] and cobalt oxides [8] have been used for the OER in previous studies, with multiple reports on bimetallic and mixed oxide systems containing cobalt listed in [9]. Multiple studies investigated cobalt-based materials for supercapacitor applications [10].

Nanocomposites that integrate carbon with cobalt present an attractive approach toward harnessing the beneficial properties of both constituents and potentially achieving synergistic effects. Multiple cobalt-doped graphitic materials were investigated with respect to their OER performance, including cobalt-embedded nitrogen-doped carbon nanotubes [11,12], as well as composites containing porous carbons [13,14]. Supercapacitor applications of cobalt oxide with graphene nanosheets were presented by Naveen et al. [15] and Lakra et al. [16]. Furthermore, Liu et al. [17] designed a composite using mesoporous carbon nanospheres decorated with cobalt-based nanoparticles that showed enhanced supercapacitor performance, demonstrating ongoing research in those topics.

The graphene-based materials discussed in the literature are synthesized using a variety of methods, such as the mechanical exfoliation of graphite [18], the reduction of graphene oxide [19] and chemical-vapor-deposition-based synthesis [20], among others [21,22]. However, purely physical techniques like plasma-assisted exfoliation of natural graphite are rarely employed. Our previous work pioneered this approach, transforming natural graphite into nanoporous few-layer graphene (FLG) with a markedly high specific surface area that was used for hydrogen storage [23], followed by a study on water purification and supercapacitor performance [24]. Hence, this method not only highlights the potential for a more efficient synthesis process but also suggests improved performance characteristics for energy-related applications.

In this work, we introduce an innovative approach by implementing an additional plasma-processing step after mixing FLG with cobalt powder, yielding cobalt-enriched few-layer graphene (Co-FLG). This composite is compared to its pure FLG counterpart through a series of structural and electrochemical characterization methods, including scanning electron microscopy (SEM), energy-dispersive X-ray spectroscopy (EDX), transmission electron microscopy (TEM), X-ray diffraction (XRD), Raman spectroscopy, gas sorption analysis (GSA) and X-ray photoelectron spectroscopy (XPS). The electrochemical analyses specifically assess the oxygen evolution reaction and supercapacitor performances. This study marks the first demonstration that a purely physical method can effectively create carbon–cobalt composite materials exhibiting enhanced electrochemical performance compared to the pure nanoporous graphene. The method stands out for its dry/chemical-free, cost-efficient and environmentally friendly character, presenting a scalable solution for industry. The findings of this study not only advance the understanding of metal–carbon composites but also open new avenues for the development of green energy technologies based on these novel materials.

2. Materials and Methods

2.1. Material Synthesis

The FLG material was synthesized by subjecting natural flake graphite (Asbury Carbons, Ashbury, NJ) to plasma treatment using a multi-electrode dielectric barrier discharge plasma reactor, operating under the conditions outlined in [23]. The graphite was subjected to an argon flow of 600 sccm (corresponding to a pressure of 0.1 mbar) with a power of 6 kW for 60 min. Following this, the resultant FLG powder underwent a further 30 min oxygen plasma treatment to add oxygen-containing functional groups aimed at decreasing particle agglomeration.

A total of 100 g of the FLG material was then mixed with 5 g of cobalt powder (Thermo Scientific Chemicals, Waltham, MA, USA) of 99.8% purity and a 1.6 µm average particle size, using mechanical stirring for 5 min to ensure uniformity. This mixture was then subjected to an additional argon plasma treatment step for 60 min within the same reactor using the same gas flow and power values as above. These specific conditions/process parameters were determined in preliminary studies and were chosen to promote the uniform distribution of cobalt particles within the FLG matrix and to optimize the composite's physical and electrochemical characteristics.

2.2. Structural and Morphological Characterization

The SEM micrographs were collected by an FEI Quanta 200 microscope (Hillsboro, OR, USA) using a 20 kV acceleration voltage and a working distance of 10 mm. EDX elemental maps were collected with an EDVAC Genesis X-ray analysis probe mounted on the SEM instrument using the same 20 kV acceleration voltage.

The TEM micrographs were recorded by employing a Philips CM-20 (Eindhoven, Netherlands) transmission electron microscope with high-resolution capabilities equipped with a LaB_6 filament and operated at 200 kV acceleration voltage. The powders were first ultra-sonicated in ethanol and then placed onto holey carbon-only support films mounted on copper grids.

The XRD investigations were performed on a Bruker-AXS D8 Advance diffractometer (Karlsruhe, Germany) in a Bragg–Brentano configuration. The diffractograms were recorded using Cu Kα radiation with a wavelength of ~0.154 nm operating with a voltage of 40 kV and a current of 40 mA. The scans were recorded from diffraction angles of 15°–60° in a continuous scan mode with a step size of 0.02° and a time of 1.2 s per step. Bragg's law was used to calculate the interplanar spacing d between the individual graphene layers from the (002) reflection. Scherrer's equation was used to calculate the crystallite size L_c from the (002) reflection [25]. The number of graphene layers n was estimated by dividing the L_c value by the d value ($n = L_c/d$).

Micro-Raman spectroscopy was conducted utilizing an inVia Reflex spectrometer (Renishaw, Wotton-under-Edge, UK), equipped with a 514.5 nm solid-state laser for excitation. The laser beam, focused to a 4.5 mm diameter spot, was directed through an objective lens featuring ×20 magnification and a numerical aperture of 0.14. A power density of 0.02 W/μm^2, following guidelines from a prior study [26], was applied.

The GSA studies were performed on an Autosorb iQ^3 gas sorption analyzer (Anton Paar QuantaTec, Boynton Beach, FL, USA) using high-purity (99.999%) N_2 and He gases for the adsorption measurements and void volume calculations, respectively, with liquid N_2 at 77 K as a cryogen. The samples (~50 mg) were degassed at 250 °C under vacuum for 24 h prior to performing the GSA measurements. To determine the specific surface area, the multi-point Brunauer–Emmett–Teller (BET) method [27] was utilized, adhering to the BET consistency criteria outlined in ISO 9277:2022. This calculation was based on adsorption data at relative pressures (P/P_0) ranging from 0.01 to 0.05. The values for the micropore surface area and micropore volume were calculated using the t-plot method for carbon black in the relative pressure (P/P_0) interval from 0.2 to 0.5. The pore size distributions were estimated using the quenched solid density functional theory (QSDFT). This method employed the N_2–carbon adsorption branch kernel at 77 K for analyzing mixed slit and cylindrical pores.

The XPS analysis was conducted on a Theta Probe system (Thermo Scientific, Waltham, MA, USA) featuring a monochromated Al Kα X-ray source with 1486.6 eV photon energy. The X-ray spot was focused to approximately 400 µm in diameter. The wide-scan survey spectra were collected at a pass energy of 200 eV, and the high-resolution core level spectra for various elements were obtained with a pass energy of 50 eV.

2.3. Electrochemical Characterization

The electrochemical properties were assessed using a three-electrode configuration in a 1 M KOH solution (pH value 13.89–14.06 [28]) for a catalytic activity analysis (using OER) and a 3 M KOH solution (pH value 14.56–14.74 [28]) for the energy storage studies (using supercapacitors). The setup consisted of a saturated calomel electrode as the reference, a platinum wire as the counter and custom-made electrodes as the working electrodes. These electrodes were fabricated by blending 80 wt.% FLG or Co-FLG with 10wt.% acetylene black and 10 wt.% polyvinylidene difluoride in N-methyl pyrrolidinone, applied onto nickel foam substrates with a total working electrode area of 0.25 cm^2. For the OER studies, linear sweep voltammetry (LSV) and a Tafel slope analysis were employed to assess the kinetics and efficiency of the catalytic process. For evaluating the supercapacitive behavior, cyclic voltammetry (CV) and long-term durability tests, extending over 5000 cycles, were performed to assess the charge storage efficiency and electrode stability. These tests were conducted using a VersaSTAT 4-500 electrochemical workstation (Princeton Applied Research, Oak Ridge, TN, USA), allowing for a comprehensive analysis of both the catalytic activity and energy storage capabilities.

3. Results and Discussion

3.1. Structural and Morphological Investigations

The field-emission SEM micrographs of both the FLG and Co-FLG samples reveal a mix of individual spherical particle agglomerates, with sizes typically below 1 μm in diameter, at lower magnifications (Figure 1a,d). At higher magnifications (Figure 1b,e), a platelet-like structure is visible. The high-resolution TEM micrographs of the FLG (Figure 1c) reveal individual thin flakes with various structural defects. In contrast, the Co-FLG TEM micrographs (Figure 1f) display a mixture of carbon flakes and spherical particles, the latter appearing darker and indicating cobalt's presence. The EDX analysis of the FLG identifies a high oxygen content alongside carbon, indicative of successful functionalization. For the Co-FLG sample, EDX confirms the presence of cobalt. The EDX mappings of the spherical agglomerates were conducted for carbon (Figure 1g), oxygen (Figure 1h) and cobalt (Figure 1i), revealing that carbon is uniformly distributed throughout the sample, while cobalt is significantly less predominant and concentrated along specific surface sites of the sample.

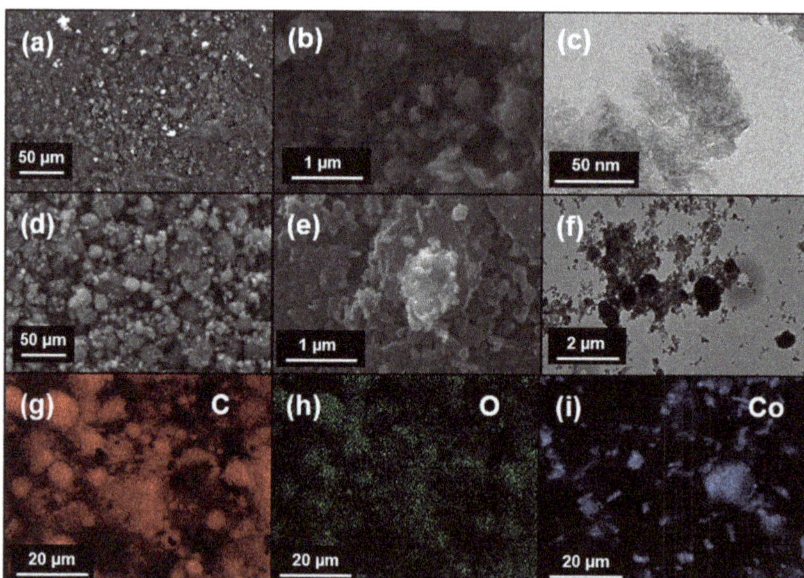

Figure 1. (**a**,**b**) SEM and (**c**) TEM micrographs for FLG. (**d**,**e**) SEM and (**f**) TEM micrographs for Co-FLG. EDX mappings of Co-FLG for (**g**) carbon, (**h**) oxygen and (**i**) cobalt.

XRD experiments were conducted, with the diffractograms depicted in Figure 2a. The FLG sample reveals a distinct (002) peak at a diffraction angle 2θ of 26.3° and a broad (100) peak located around 43.3°, both associated with the graphitic structure (International

Centre for Diffraction Data card no. 75-1621). The Co-FLG sample reveals graphitic peaks as well as strong contributions attributed to both the face-centered cubic (fcc) and hexagonal close-packed (hcp) phases of cobalt. While the hcp phase is thermodynamically stable at room temperature, the two phases are known to coexist at room temperature [29]. Notably, there are no peaks indicating the presence of cobalt oxides (i.e., CoO and Co_3O_4) in the XRD patterns. Applying Bragg's law, the interplanar spacing d between the individual graphene layers is calculated as 0.338 nm for both samples, slightly higher than the theoretical graphite interlayer spacing of 0.335 nm [30]. Applying Scherrer's equation allows for calculating the crystallite size L_c of 5.7 and 5.4 nm and correspondingly an average number of layers of ~17 and 16 layers for the FLG and Co-FLG, respectively. The striking similarities between both samples imply that the additional plasma treatment step does not cause significant structural changes to the carbon phase of the Co-FLG sample.

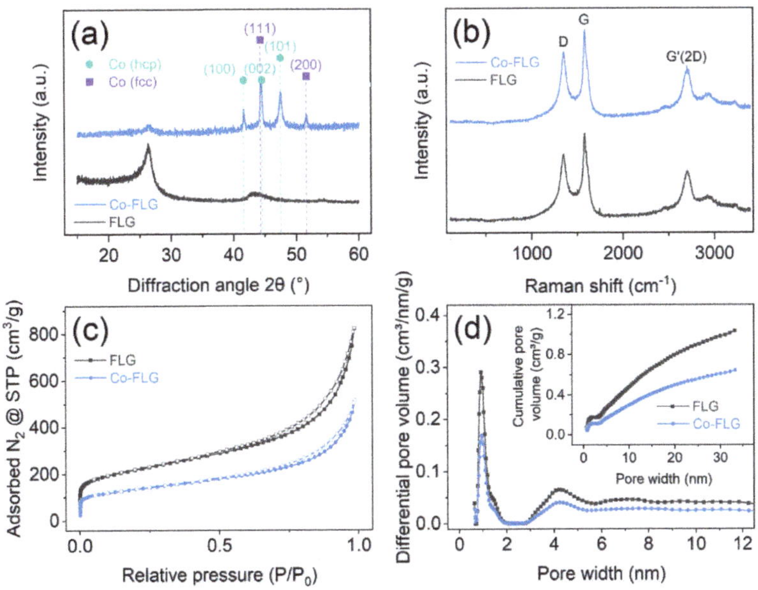

Figure 2. Results of structural characterization of FLG (black) and Co-FLG (blue): (**a**) X-ray diffractograms with the positions of the main diffraction peaks of the allotropic fcc and hcp phases of cobalt and (**b**) Raman spectra with indications for the D, G and G'(2D) bands. (**c**) Gas adsorption/desorption isotherms for N_2 at 77 K and corresponding (**d**) QSDFT-derived pore size distributions with an inset showing the cumulative pore volume.

Raman spectroscopy was conducted and the results are shown in Figure 2b, revealing three distinct features known as the D, G and G'(2D) bands of carbons [31] located at 1353, 1576 and 2695 cm^{-1}, respectively. All the peak positions and integrated peak areas are virtually identical for both samples, with the presence of a strong D band suggesting a high defect density in the carbon structure. The shape and position of the G'(2D) band indicates a graphene-based structure with a few layers, rather than graphite [32]. The extensive similarities between both Raman spectra suggest that the incorporation of cobalt does not alter the carbon structure. The anticipated peaks for the cobalt oxides, which would typically appear at 690 cm^{-1} for Co_3O_4 [33] and a broad band around 1060 cm^{-1} for CoO [34,35], are absent, supporting the findings from the XRD. The defect density is calculated for both samples as ~1.88×10^{11} cm^{-2} following a procedure outlined by Cançado et al. [36]. The results of this investigation show that no significant structural evolution of the nanoporous graphene phase was found after conducting the second plasma treatment step.

The results of the gas sorption analysis are depicted in Figure 2c,d. Figure 2c shows the gas adsorption (filled symbols) and desorption (empty symbols) isotherms using N_2 at 77 K. The sharp increase at low relative pressures can be attributed to microporosity (i.e., pore widths below 2 nm). The isotherms are similar in shape, with the FLG sample displaying a significantly higher N_2 adsorption amount throughout the full pressure range. For the concluding point of the adsorption isotherms (P/P_0 ~0.99), the Co-FLG exhibits a reduced N_2 adsorption of roughly 37% compared to the FLG. The desorption branch for both samples shows a small hysteresis, indicating the presence of mesopores (i.e., pore widths between 2 and 50 nm). The adsorption isotherms do not end with a plateau; thus, the Gurvich rule cannot be applied to determine the total pore volume [37]. The isotherms are classified as a mixed Type I and Type IV(a) according to the classification of the International Union of Pure and Applied Chemistry, as both a contribution from the micropores in the low pressure range and the mesopores in the intermediate pressure range is visible [38]. A small step-down in the desorption branch in the P/P_0 range between 0.5 and 0.4 is visible for both samples. This is commonly reported in the literature and is attributed to cavitation-induced N_2 evaporation from the mesopores [37,39]. Thus, the adsorption branch is used for extracting the pore size distribution depicted in Figure 2d. Using the desorption branch would yield an artifact in the pore size distribution in the range between 3 and 4 nm, associated with the cavitation-induced evaporation mechanism [39]. The pore size distribution for both samples is virtually identical, exhibiting a bimodal distribution with peaks at 0.93 and 4.22 nm, thus revealing super-micropores (i.e., pore widths between 0.7 and 2 nm) and small mesopores, respectively. It seems that there are no pores in the size range of 2 to 2.8 nm present within the investigated samples. The inset in Figure 2d reveals the cumulative pore size distribution for both samples. The BET method reveals apparent surface areas of 780 m^2/g for the FLG and 484 m^2/g for the Co-FLG, which corresponds to a reduction of roughly 38% upon cobalt addition. The t-plot method was employed to extract the micropore surface area and the micropore volume of the given samples. The overall contribution of the micropores is roughly 35% for both samples. The remaining surface area S_{ext} consists of contributions from mesopores, macropores and the external surface. However, a more precise quantification cannot be performed, as the isotherm lacks a plateau in the high-pressure range. Consequently, no mean pore width value can be reported. The results of the GSA are summarized in Table 1. To conclude, both samples show a virtually identical pore size distribution, while Co-FLG displayed a decrease of roughly 38% with respect to both the specific surface area and the micropore volume. This effect is attributed to pore blocking, as previously described in literature [26].

Table 1. Results of the gas sorption analysis conducted using N_2 at 77 K.

Material	S_{BET} [m^2/g]	S_{micro} [m^2/g]	S_{ext} [m^2/g]	V_{micro} [cm^3/g]
FLG	780	279	501	0.123
Co-FLG	484	175	309	0.077
Change	−37.9%	−37.3%	−38.4%	−37.4%

S_{BET}: Brunauer–Emmett–Teller (BET) specific surface area. **S_{micro}**: Specific surface area calculated by the statistical thickness t-plot method. **S_{ext}**: Specific surface area contributed by mesopores, macropores and external surfaces, calculated as difference between S_{BET} and S_{micro}. **V_{micro}**: Specific micropore volume calculated by the t-plot method.

XPS measurements were performed to analyze the elemental composition of the samples. Figure 3a presents the wide-scan survey spectra, showing carbon and oxygen in both samples and additionally cobalt in the Co-FLG. A quantitative analysis from the survey scans determines the chemical compositions to be 94.1 and 91.9 at.% of carbon, 5.6 and 6.7 at.% of oxygen, small quantities of nitrogen (0.3 and 0.6 at.%) for the FLG and Co-FLG, respectively, and 0.8 at.% of cobalt for the Co-FLG. This translates to 3.3 wt.% of cobalt, which is slightly lower than the anticipated 4.8 wt.% based on the synthesis procedure.

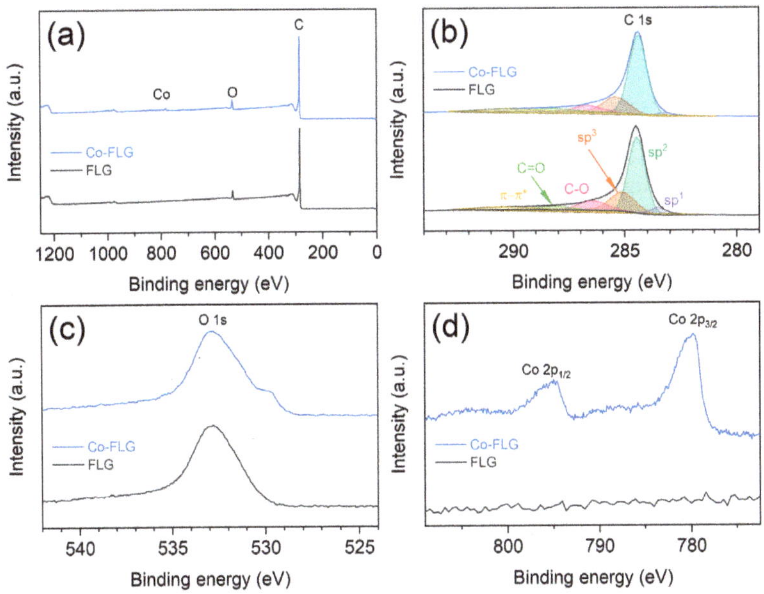

Figure 3. Results of XPS investigations on FLG (black) and Co-FLG (blue) consisting of a (**a**) survey scan and high-resolution spectra for the elements (**b**) carbon, (**c**) oxygen and (**d**) cobalt.

Figure 3b displays the high-resolution spectra of carbon, where the peak deconvolution reveals the presence of sp^1, sp^2, sp^3, C-O and C=O bonds, along with a π-π^* satellite peak in the 290–292 eV region [40]. The spectra are predominantly characterized by sp^2-hybridized carbon bonds associated with the peak position at 284.5 eV, which represent the largest fraction of the carbon content, accounting for ~56% in both examined samples. Contributions from sp^3-hybridized carbon are also significant, comprising ~21% and ~16% for the FLG and Co-FLG samples, respectively, with small contributions from the remaining carbon bonds.

High-resolution oxygen spectra are depicted in Figure 3c, with a peak shape indicative of the presence of OH^- species and adsorbed H_2O. The additional peak at low binding energies corresponds to cobalt oxide formation [41]. Figure 3d confirms cobalt's presence in the Co-FLG, in good agreement with the EDX-derived elemental map for cobalt (see Figure 1i). The Co $2p_{3/2}$ peak occurs at a binding energy of 779.8 eV. This binding energy together with only a weak satellite structure at around 790 eV between the $2p_{3/2}$ and $2p_{1/2}$ peaks is characteristic of Co_3O_4, [41–43], as expected for the native oxide [41]. Based on the theoretical work, cobalt and cobalt oxides are expected to transform into CoOOH under the conditions present during an OER [44].

The incorporation of cobalt into the FLG, as indicated by Figures 1i, 2a and 3d, leaves the carbon structure unaltered, as evidenced by the Raman spectroscopy and gas sorption analyses. The TEM observations reveal mostly spherical cobalt particles, with their presence further validated by EDX mapping, XRD and XPS. The overall pore size distribution remains consistent despite the cobalt addition. However, a notable reduction in the specific surface area by 38% is observed. The incorporation of cobalt might enable tuning the material's functionality. In the next step, the electrochemical performance differences between FLG and Co-FLG were investigated by means of OER and supercapacitor energy storage applications.

3.2. Electrochemical Characterization

The electrochemical performance of FLG and Co-FLG in terms of the OER was analyzed using LSV, as shown in Figure 4a. As the potential increases, the current density for both materials also increases, which is typical for catalytic processes, as the overpotential provides additional driving force for the reactions. Notably, the Co-FLG exhibits a higher current density than the FLG at potentials beyond approximately 1.55 V, suggesting that cobalt doping enhances the electrocatalytic activity of the graphene material. At a benchmark current density of 10 mA/cm^2, overpotentials of 280 mV for FLG and 290 mV for Co-FLG are observed, comparable to the 300 mV/cm^2 reported by Dou et al. [13] for a porous carbon doped with cobalt nanoparticles in 0.1 M KOH media and mirroring the 290 mV/cm^2 reported by Jia et al. [45] for cobalt nanoparticles embedded in nitrogen-doped carbon on carbon cloth in a 1 M KOH solution.

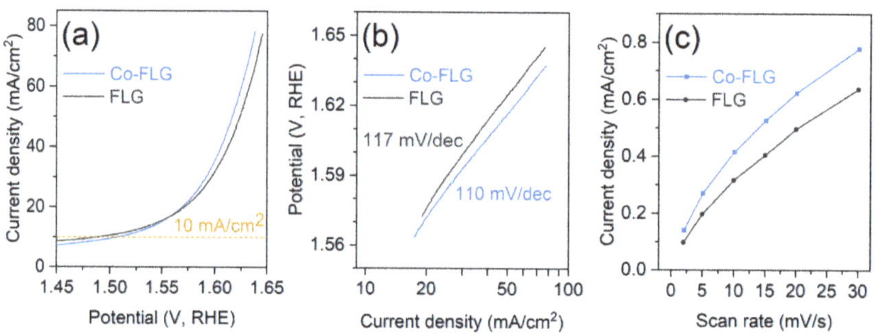

Figure 4. Characterization of OER performance. (**a**) LSV curves, (**b**) corresponding Tafel slopes and (**c**) current density as a function of the scan rate.

A further analysis of the OER kinetics reveals that Co-FLG has a lower Tafel slope (110 mV/dec) compared to FLG (117 mV/dec), as seen in Figure 4b. This suggests that Co-FLG requires a smaller increase in potential to achieve a tenfold increase in current density compared to FLG, which indicates that Co-FLG has a better catalytic efficiency for the OER. The lower Tafel slope of the Co-FLG sample indicates a higher electrocatalytic surface area. The Tafel slope of 117 mV/dec for the FLG sample implies that an electrochemical step is determining the rate-limiting step rather than chemical reactions based on the theoretical works [46,47]. For the case of Co-FLG, the reported Tafel slope of 110 mV/dec is not directly covered by the theoretical calculations in the literature, which also applies for subsequent experimentally derived Tafel slope values. Jia et al. reported various Tafel slopes of 73 and 88 mV/dec for two different types of N-doped carbon cloth materials also doped with cobalt and 253 mV/dec for the N-doped carbon cloth without cobalt addition under a 1 M KOH solution. The synthesis procedure consisted of electrochemically deposited Co_3O_4 nanosheets on the carbon cloth surface [45]. Dou et al. reported a significantly lower Tafel slope of 96.9 mV/dec for a cobalt nanoparticle-doped porous carbon and a lower Tafel slope of 73 mV/dec for a more advanced material system incorporating carbon nanotubes [13], however with no reasons given for the differences in the Tafel slopes. It has to be noted that the cobalt nanoparticles were significantly smaller compared to this work and a 0.1 M KOH solution was used. Mao et al. reported on 3D crumpled graphene–cobalt oxide nanohybrids with an overpotential of 340 mV at 10 mA/cm^2 and Tafel slopes of 71 and 75 mV/dec for N-doped and non-doped versions of the material system [48]. Su et al. reported Tafel slopes between 61.4 and 116.1 mV/dec for cobalt nanoparticles embedded in N-doped carbon with values depending on different synthesis parameters of the material system [49].

The current density was studied as a function of the scan rate (see Figure 4c). Co-FLG is characterized by consistently higher current densities compared to FLG, culminating in a

22% higher current density at the highest tested scan rate of 30 mV/s. This enhancement in electrochemical activity underscores Co-FLG's potential for high-current-density applications.

To summarize, the presented material system is capable of achieving a competitive overpotential at 10 mA/cm^2 while only obtaining intermediate Tafel slopes. The material system could be further enhanced by N-doping the few-layer graphene [3]. This could be performed within the plasma reactor using an additional plasma treatment step with nitrogen gas.

Figure 5a presents a comparative evaluation of the specific capacitance as a function of the current density for FLG and Co-FLG in supercapacitor applications. Both curves exhibit a decrease in specific capacitance with increasing current density. At a current density of 0.5 A/g, FLG achieves a significantly higher specific capacitance (125.8 F/g) compared to Co-FLG (99.6 F/g), marking a 26% increase. The specific capacitance of FLG decreases significantly with the current density, achieving roughly equal values of 62 F/g at 3 A/g compared to Co-FLG and terminating with 3.6 F/g at the highest measured current density of 20 A/g. In contrast, Co-FLG retains a significantly higher specific capacitance at elevated current densities, i.e., 24.6 F/g at 20 A/g, thus outperforming the conventional FLG by a factor of 6.8. The results for the specific capacitance as a function of the scan rate are depicted in Figure 5b. In a similar fashion to Figure 5a, Co-FLG starts at a lower specific capacitance at low scan rates and outperforms the FLG at higher scan rates, despite a less pronounced difference of 27%. This indicates that Co-FLG might be particularly useful for applications requiring rapid charge–discharge cycles, where maintaining the energy storage capacity is crucial. Benchmarking the performance of supercapacitors is complex, as standardized benchmarks, similar to those for the OER, are absent. Lu et al. addressed this complexity in their recent review, showcasing a range of cobalt-based graphene materials [50]. The Co_3O_4/graphene nanosheets composite prepared by Naveen et al. performed significantly better with 175 F/g at 10 A/g in a 1 M KOH solution (compared to 37.9 F/g for Co-FLG) and with 650 F/g at 5 mV/s (compared to 283 F/g for Co-FLG) in a 1 M KOH solution [15]. Dong et al. reported on a 3D graphene foam–cobalt oxide nanowire electrode achieving 768 F/g at a current density of 10 A/g (compared to 38 F/g for Co-FLG) combined with outstanding cycling stability [51]. Yan et al. reported on a similar nanocomposite synthesized using a plasma-assisted process, achieving 243 F/g at 5 mV/s (compared to 236.7 F/g for Co-FLG) in a 6 M KOH aqueous solution [52].

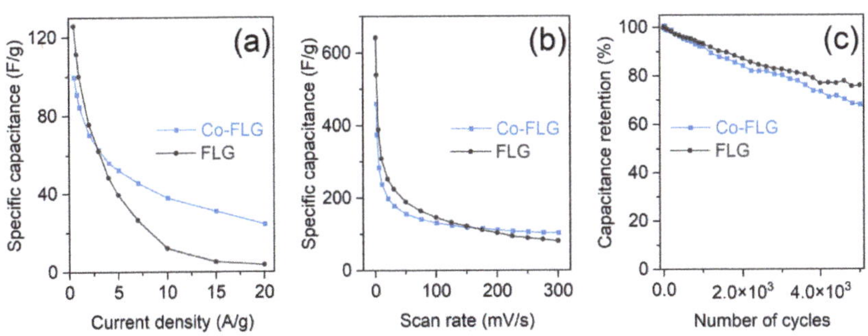

Figure 5. Supercapacitor performance. (**a**) Specific capacitance as a function of the current, (**b**) specific capacitance as a function of the scan rate and (**c**) capacitance retention over 5000 cycles.

When the specific capacitance is normalized by the BET specific surface area, yielding values in F/m^2, Co-FLG consistently outperforms FLG. At a current density of 0.5 A/g, Co-FLG achieves a notable specific capacitance of ~0.21 F/m^2, compared to ~0.16 F/m^2 for FLG. This performance gap widens with increasing current densities; at the highest tested current density of 20 A/g, Co-FLG exceeds FLG by a factor of 11, demonstrating specific capacitances of ~0.051 and ~0.005 F/m^2, respectively. This underlines the enhanced electrochemical performance achieved by the addition of cobalt despite the reduction in the available surface area.

The cycling stability tests (Figure 5c) reveal a gradual decline in capacitance retention for both Co-FLG and FLG over 5000 cycles. FLG performs more favorably, retaining 76% of its original capacitance, while Co-FLG retains 68% after 5000 cycles. The lower capacitance retention of Co-FLG indicates more pronounced degradation from cycling, potentially due to microstructural changes, or the instability of cobalt dopants. Yan et al. reported a slight increase in cyclic capacitance for their nanocomposite within the first 250 cycles and a subsequent decrease to 95% after a total of 2000 cycles, compared to 84% for Co-FLG and 87% for FLG [52]. Vilian et al. reported a retention of 80.5% after 5000 cycles for hexagonal Co_3O_4-anchored reduced graphene oxide sheets [53]. Liu et al. reported an excellent cycling retention of 96% after 5500 cycles for a material system consisting of hierarchical porous carbon frameworks containing ultrafine metallic cobalt nanoparticles [17]. Naveen et al. reported a cycling retention of 92% after 1000 cycles, matching the performance of Co-FLG [15]. Assessing cycling stability suffers from non-uniform testing conditions and the reasons for degradation are rarely discussed within the literature.

To summarize, Co-FLG enhances the supercapacitor performance of the given FLG material and delivers reasonable performance while falling short of matching the best currently available materials. Further optimization of the sample morphology and the length scale of the cobalt phase is suggested to further enhance supercapacitor performance [50]. It has to be noted that the presented performance was achieved by means of physical methods only, thus allowing for a wet-free synthesis procedure, with no comparable results yet published to the best of the authors' knowledge.

4. Conclusions

Within this work, we pioneered the synthesis of a nanocomposite of cobalt and nanoporous graphene synthesized by purely physical plasma treatment methods without the need for chemicals and cobalt precursors. We demonstrated for the first time that such processes can be successfully employed to obtain innovative material systems with a competitive oxygen evolution reaction and supercapacitor performance while retaining the advantageous properties of nanoporous graphene. However, the presented method does not yet allow for achieving an electrochemical performance on the same level as the best cobalt–carbon nanocomposites available in the literature. Looking forward, the incorporation of nitrogen into the carbon structure via plasma treatment processes emerges as a potential area for further research, promising to further optimize electrochemical performance.

Author Contributions: Conceptualization, F.K. and N.K.; methodology, F.K. and N.K.; formal analysis, F.K., N.K., R.K.G., S.H. and M.B.; investigation, F.K., N.K., R.K.G., A.T. and S.H.; resources, N.K., R.K.G., A.T. and C.R.; writing—original draft preparation, F.K.; writing—review and editing, N.K., R.K.G., A.T., S.H., M.B., C.R. and C.M.; visualization, F.K.; supervision, C.M. All authors have read and agreed to the published version of the manuscript.

Funding: This research received no external funding.

Data Availability Statement: The data that support the findings of this study are available from the corresponding authors upon reasonable request.

Acknowledgments: The authors are grateful to Konstantinos Giannakopoulos from NCSR Demokritos (Greece) for the TEM investigations, Georgios Constantinides from the Cyprus University of Technology (Cyprus) for the SEM/EDX investigations and Athanassios G. Kontos from the National Technical University of Athens (Greece) for the Raman spectroscopy investigations.

Conflicts of Interest: Author Afshin Tarat was employed by the company Loginns GmbH. The remaining authors declare that the research was conducted in the absence of any commercial or financial relationships that could be construed as a potential conflict of interest.

References

1. Jiao, Y.; Zheng, Y.; Jaroniec, M.; Qiao, S.Z. Design of electrocatalysts for oxygen- and hydrogen-involving energy conversion reactions. *Chem. Soc. Rev.* **2015**, *44*, 2060–2086. [CrossRef] [PubMed]
2. Bueno, P.R. Nanoscale origins of super-capacitance phenomena. *J. Power Sources* **2019**, *414*, 420–434. [CrossRef]
3. Zhang, L.; Xiao, J.; Wang, H.; Shao, M. Carbon-Based Electrocatalysts for Hydrogen and Oxygen Evolution Reactions. *ACS Catal.* **2017**, *7*, 7855–7865. [CrossRef]
4. Pandolfo, A.G.; Hollenkamp, A.F. Carbon properties and their role in supercapacitors. *J. Power Sources* **2006**, *157*, 11–27. [CrossRef]
5. Zhai, Z.; Zhang, L.; Du, T.; Ren, B.; Xu, Y.; Wang, S.; Miao, J.; Liu, Z. A review of carbon materials for supercapacitors. *Mater. Des.* **2022**, *221*, 111017. [CrossRef]
6. Wen, Z.; Wang, X.; Mao, S.; Bo, Z.; Kim, H.; Cui, S.; Lu, G.; Feng, X.; Chen, J. Crumpled Nitrogen-Doped Graphene Nanosheets with Ultrahigh Pore Volume for High-Performance Supercapacitor. *Adv. Mater.* **2012**, *24*, 5610–5616. [CrossRef] [PubMed]
7. Wu, L.; Li, Q.; Wu, C.H.; Zhu, H.; Mendoza-Garcia, A.; Shen, B.; Guo, J.; Sun, S. Stable Cobalt Nanoparticles and Their Monolayer Array as an Efficient Electrocatalyst for Oxygen Evolution Reaction. *J. Am. Chem. Soc.* **2015**, *137*, 7071–7074. [CrossRef]
8. Zhang, Y.; Ouyang, B.; Xu, J.; Jia, G.; Chen, S.; Rawat, R.S.; Fan, H.J. Rapid Synthesis of Cobalt Nitride Nanowires: Highly Efficient and Low-Cost Catalysts for Oxygen Evolution. *Angew. Chem. Int. Ed.* **2016**, *55*, 8670–8674. [CrossRef] [PubMed]
9. Tahir, M.; Pan, L.; Idrees, F.; Zhang, X.; Wang, L.; Zou, J.-J.; Wang, Z.L. Electrocatalytic oxygen evolution reaction for energy conversion and storage: A comprehensive review. *Nano Energy* **2017**, *37*, 136–157. [CrossRef]
10. Hu, X.; Wang, Y.; Wu, Q.; Li, J. Review of cobalt-based nanocomposites as electrode for supercapacitor application. *Ionics* **2022**, *28*, 989–1015. [CrossRef]
11. Qiao, X.; Liao, S.; Zheng, R.; Deng, Y.; Song, H.; Du, L. Cobalt and Nitrogen Codoped Graphene with Inserted Carbon Nanospheres as an Efficient Bifunctional Electrocatalyst for Oxygen Reduction and Evolution. *ACS Sustain. Chem. Eng.* **2016**, *4*, 4131–4136. [CrossRef]
12. Wang, Z.; Xiao, S.; Zhu, Z.; Long, X.; Zheng, X.; Lu, X.; Yang, S. Cobalt-Embedded Nitrogen Doped Carbon Nanotubes: A Bifunctional Catalyst for Oxygen Electrode Reactions in a Wide pH Range. *ACS Appl. Mater. Interfaces* **2015**, *7*, 4048–4055. [CrossRef] [PubMed]
13. Dou, S.; Li, X.; Tao, L.; Huo, J.; Wang, S. Cobalt nanoparticle-embedded carbon nanotube/porous carbon hybrid derived from MOF-encapsulated Co_3O_4 for oxygen electrocatalysis. *Chem. Commun.* **2016**, *52*, 9727–9730. [CrossRef] [PubMed]
14. Li, X.; Zeng, C.; Jiang, J.; Ai, L. Magnetic cobalt nanoparticles embedded in hierarchically porous nitrogen-doped carbon frameworks for highly efficient and well-recyclable catalysis. *J. Mater. Chem. A* **2016**, *4*, 7476–7482. [CrossRef]
15. Naveen, A.N.; Manimaran, P.; Selladurai, S. Cobalt oxide (Co_3O_4)/graphene nanosheets (GNS) composite prepared by novel route for supercapacitor application. *J. Mater. Sci. Mater. Electron.* **2015**, *26*, 8988–9000. [CrossRef]
16. Lakra, R.; Kumar, R.; Sahoo, P.K.; Sharma, D.; Thatoi, D.; Soam, A. Facile synthesis of cobalt oxide and graphene nanosheets nanocomposite for aqueous supercapacitor application. *Carbon Trends* **2022**, *7*, 100144. [CrossRef]
17. Liu, B.; Jin, L.; Zheng, H.; Yao, H.; Wu, Y.; Lopes, A.; He, J. Ultrafine Co-based Nanoparticle@Mesoporous Carbon Nanospheres toward High-Performance Supercapacitors. *ACS Appl. Mater. Interfaces* **2017**, *9*, 1746–1758. [CrossRef] [PubMed]
18. Yi, M.; Shen, Z. A review on mechanical exfoliation for the scalable production of graphene. *J. Mater. Chem. A* **2015**, *3*, 11700–11715. [CrossRef]
19. Pei, S.; Cheng, H.-M. The reduction of graphene oxide. *Carbon* **2012**, *50*, 3210–3228. [CrossRef]
20. Tetlow, H.; Posthuma de Boer, J.; Ford, I.J.; Vvedensky, D.D.; Coraux, J.; Kantorovich, L. Growth of epitaxial graphene: Theory and experiment. *Phys. Rep.* **2014**, *542*, 195–295. [CrossRef]
21. Mohan, V.B.; Lau, K.; Hui, D.; Bhattacharyya, D. Graphene-based materials and their composites: A review on production, applications and product limitations. *Compos. Part B Eng.* **2018**, *142*, 200–220. [CrossRef]
22. Mbayachi, V.B.; Ndayiragije, E.; Sammani, T.; Taj, S.; Mbuta, E.R.; Khan, A. ullah Graphene synthesis, characterization and its applications: A review. *Results Chem.* **2021**, *3*, 100163. [CrossRef]
23. Kostoglou, N.; Tarat, A.; Walters, I.; Ryzhkov, V.; Tampaxis, C.; Charalambopoulou, G.; Steriotis, T.; Mitterer, C.; Rebholz, C. Few-layer graphene-like flakes derived by plasma treatment: A potential material for hydrogen adsorption and storage. *Microporous Mesoporous Mater.* **2016**, *225*, 482–487. [CrossRef]
24. Natter, N.; Kostoglou, N.; Koczwara, C.; Tampaxis, C.; Steriotis, T.; Gupta, R.; Paris, O.; Rebholz, C.; Mitterer, C. Plasma-Derived Graphene-Based Materials for Water Purification and Energy Storage. *C* **2019**, *5*, 16. [CrossRef]
25. Warren, B.E. X-ray Diffraction in Random Layer Lattices. *Phys. Rev.* **1941**, *59*, 693–698. [CrossRef]
26. Kostoglou, N.; Liao, C.W.; Wang, C.Y.; Kondo, J.N.; Tampaxis, C.; Steriotis, T.; Giannakopoulos, K.; Kontos, A.G.; Hinder, S.; Baker, M.; et al. Effect of Pt nanoparticle decoration on the H2 storage performance of plasma-derived nanoporous graphene. *Carbon* **2021**, *171*, 294–305. [CrossRef]
27. Brunauer, S.; Emmett, P.H.; Teller, E. Adsorption of Gases in Multimolecular Layers. *J. Am. Chem. Soc.* **1938**, *60*, 309–319. [CrossRef]
28. Hausmann, J.N.; Traynor, B.; Myers, R.J.; Driess, M.; Menezes, P.W. The pH of Aqueous NaOH/KOH Solutions: A Critical and Non-trivial Parameter for Electrocatalysis. *ACS Energy Lett.* **2021**, *6*, 3567–3571. [CrossRef]
29. Lee, B.W.; Alsenz, R.; Ignatiev, A.; Van Hove, M.A. Surface structures of the two allotropic phases of cobalt. *Phys. Rev. B* **1978**, *17*, 1510–1520. [CrossRef]

30. Delhaes, P. *Graphite and Precursors*; Delhaes, P., Ed.; Gordon & Breach Science Publishers: Langhorne, PA, USA, 2001; Volume 1, ISBN 9056992287.
31. Cuesta, A.; Dhamelincourt, P.; Laureyns, J.; Martinez-Alonso, A.; Tascón, J.M.D. Raman microprobe studies on carbon materials. *Carbon* **1994**, *32*, 1523–1532. [CrossRef]
32. Ferrari, A.C.; Meyer, J.C.; Scardaci, V.; Casiraghi, C.; Lazzeri, M.; Mauri, F.; Piscanec, S.; Jiang, D.; Novoselov, K.S.; Roth, S.; et al. Raman Spectrum of Graphene and Graphene Layers. *Phys. Rev. Lett.* **2006**, *97*, 187401. [CrossRef] [PubMed]
33. Hadjiev, V.G.; Iliev, M.N.; Vergilov, I. V The Raman spectra of Co_3O_4. *J. Phys. C Solid State Phys.* **1988**, *21*, L199. [CrossRef]
34. Wdowik, U.D.; Parlinski, K. Lattice dynamics of CoO from first principles. *Phys. Rev. B* **2007**, *75*, 104306. [CrossRef]
35. Li, Y.; Qiu, W.; Qin, F.; Fang, H.; Hadjiev, V.G.; Litvinov, D.; Bao, J. Identification of Cobalt Oxides with Raman Scattering and Fourier Transform Infrared Spectroscopy. *J. Phys. Chem. C* **2016**, *120*, 4511–4516. [CrossRef]
36. Cançado, L.G.; Jorio, A.; Ferreira, E.H.M.; Stavale, F.; Achete, C.A.; Capaz, R.B.; Moutinho, M.V.O.; Lombardo, A.; Kulmala, T.S.; Ferrari, A.C. Quantifying Defects in Graphene via Raman Spectroscopy at Different Excitation Energies. *Nano Lett.* **2011**, *11*, 3190–3196. [CrossRef] [PubMed]
37. Rouquerol, J.; Rouquerol, F.; Llewellyn, P.; Maurin, G.; Sing, K. *Adsorption by Powders and Porous Solids: Principles, Methodology and Applications*; Academic Press: Cambridge, MA, USA, 2013; ISBN 0080970362.
38. Thommes, M.; Kaneko, K.; Neimark, A.V.; Olivier, J.P.; Rodriguez-Reinoso, F.; Rouquerol, J.; Sing, K.S.W. Physisorption of gases, with special reference to the evaluation of surface area and pore size distribution (IUPAC Technical Report). *Pure Appl. Chem.* **2015**, *87*, 1051–1069. [CrossRef]
39. Thommes, M.; Cychosz, K.A. Physical adsorption characterization of nanoporous materials: Progress and challenges. *Adsorption* **2014**, *20*, 233–250. [CrossRef]
40. Rybachuk, M.; Bell, J.M. Electronic states of trans-polyacetylene, poly(p-phenylene vinylene) and sp-hybridised carbon species in amorphous hydrogenated carbon probed by resonant Raman scattering. *Carbon* **2009**, *47*, 2481–2490. [CrossRef]
41. McIntyre, N.S.; Johnston, D.D.; Coatsworth, L.L.; Davidson, R.D.; Brown, J.R. X-ray photoelectron spectroscopic studies of thin film oxides of cobalt and molybdenum. *Surf. Interface Anal.* **1990**, *15*, 265–272. [CrossRef]
42. National Institute of Standards and Technology, NIST X-ray Photoelectron Spectroscopy Database (SRD 20), Version 5.0. Available online: https://srdata.nist.gov/xps/ (accessed on 15 February 2024).
43. Vaz, C.A.F.; Prabhakaran, D.; Altman, E.I.; Henrich, V.E. Experimental study of the interfacial cobalt oxide in $Co_3O_4/\alpha-Al_2O_3$(0001) epitaxial films. *Phys. Rev. B* **2009**, *80*, 155457. [CrossRef]
44. Bajdich, M.; García-Mota, M.; Vojvodic, A.; Nørskov, J.K.; Bell, A.T. Theoretical Investigation of the Activity of Cobalt Oxides for the Electrochemical Oxidation of Water. *J. Am. Chem. Soc.* **2013**, *135*, 13521–13530. [CrossRef] [PubMed]
45. Jia, Q.; Gao, Y.; Li, Y.; Fan, X.; Zhang, F.; Zhang, G.; Peng, W.; Wang, S. Cobalt nanoparticles embedded in N-doped carbon on carbon cloth as free-standing electrodes for electrochemically-assisted catalytic oxidation of phenol and overall water splitting. *Carbon* **2019**, *155*, 287–297. [CrossRef]
46. Fletcher, S. Tafel slopes from first principles. *J. Solid State Electrochem.* **2009**, *13*, 537–549. [CrossRef]
47. Antipin, D.; Risch, M. Calculation of the Tafel slope and reaction order of the oxygen evolution reaction between pH 12 and pH 14 for the adsorbate mechanism. *Electrochem. Sci. Adv.* **2023**, *3*, e2100213. [CrossRef]
48. Mao, S.; Wen, Z.; Huang, T.; Hou, Y.; Chen, J. High-performance bi-functional electrocatalysts of 3D crumpled graphene–cobalt oxide nanohybrids for oxygen reduction and evolution reactions. *Energy Environ. Sci.* **2014**, *7*, 609–616. [CrossRef]
49. Su, Y.; Zhu, Y.; Jiang, H.; Shen, J.; Yang, X.; Zou, W.; Chen, J.; Li, C. Cobalt nanoparticles embedded in N-doped carbon as an efficient bifunctional electrocatalyst for oxygen reduction and evolution reactions. *Nanoscale* **2014**, *6*, 15080–15089. [CrossRef] [PubMed]
50. Lu, C.; Liu, L.; Yang, Y.; Ma, Y.; Luo, Q.; Zhu, M. Recent Progress in Co_3O_4-Based Nanomaterials for Supercapacitors. *ChemNanoMat* **2023**, *9*, e202200537. [CrossRef]
51. Dong, X.-C.; Xu, H.; Wang, X.-W.; Huang, Y.-X.; Chan-Park, M.B.; Zhang, H.; Wang, L.-H.; Huang, W.; Chen, P. 3D Graphene–Cobalt Oxide Electrode for High-Performance Supercapacitor and Enzymeless Glucose Detection. *ACS Nano* **2012**, *6*, 3206–3213. [CrossRef]
52. Yan, J.; Wei, T.; Qiao, W.; Shao, B.; Zhao, Q.; Zhang, L.; Fan, Z. Rapid microwave-assisted synthesis of graphene nanosheet/Co_3O_4 composite for supercapacitors. *Electrochim. Acta* **2010**, *55*, 6973–6978. [CrossRef]
53. Vilian, A.T.E.; Dinesh, B.; Rethinasabapathy, M.; Hwang, S.-K.; Jin, C.-S.; Huh, Y.S.; Han, Y.-K. Hexagonal Co_3O_4 anchored reduced graphene oxide sheets for high-performance supercapacitors and non-enzymatic glucose sensing. *J. Mater. Chem. A* **2018**, *6*, 14367–14379. [CrossRef]

C Journal of Carbon Research

MDPI

Article

Supercapacitor Performance of MXene-Coated Carbon Nanofiber Electrodes

Seon Kyung Kim, Seung Ah Kim, Yoon Soo Han and Kyung-Hye Jung *

Department of Advanced Materials and Chemical Engineering, Daegu Catholic University, Gyeongsan 38430, Republic of Korea; seon3193@naver.com (S.K.K.); tmddk554@naver.com (S.A.K.); yshancu@cu.ac.kr (Y.S.H.)

* Correspondence: khjung@cu.ac.kr

Abstract: MXenes consisting of thin layers of transition metal carbides or nitrides are good candidates for electrode materials due to their excellent electrical conductivity and fast ion transfer. Electrospun carbon nanofibers are highly porous and electrically conductive, making them attractive for electrode materials. In this study, free-standing electrodes were prepared by the dip-coating of carbon nanofibers (CNFs) in the MXene (Ti_3C_2) colloidal solution, which was synthesized via the wet-etching of MAX (Ti_3AlC_2) phase, and their chemical structures were investigated by X-ray diffraction and Fourier transform infrared spectroscopy. In addition, scanning and transmission electron microscopy were used to investigate the morphological and crystallographic features of MXene-coated CNFs. Surface area and pore volumes were investigated by nitrogen adsorption/desorption measurements. Supercapacitor performance was studied by assembling a 3-electrode system with 1M aqueous sodium sulfate solution as an electrolyte. MXene-coated CNFs exhibited a maximum specific capacitance of 514 F/g at 0.5 A/g, with energy and power densities of 71.4 Wh/kg at 0.5 A/g and 2.3 kW/kg at 5 A/g, respectively, which are relevantly higher compared to the pristine CNFs due to the pseudocapacitive behavior of MXenes. They also showed comparable cyclic stability during 5000 cycles with the CNFs. This result indicates that MXene-coated carbon nanofibers can be effective electrode materials for electrochemical energy storage.

Keywords: supercapacitors; pseudocapacitors; MXene; polyacrylonitrile; carbon nanofibers

1. Introduction

2D materials have gained the attention of researchers due to their large surface area and superior electrical properties, which makes them promising for use in energy storage devices [1]. Among them, transition metal carbide and/or nitrides, MXenes, have been intensively studied as 2D ceramic materials with a general formula of $M_{n+1}X_n$, where M represents transition metals and X means carbon and/or nitrogen [2]. They are typically synthesized via wet etching the parent ternary carbides and/or nitrides, MAX phases, using chemical etchants such as hydrofluoric acid (HF) [3]. The general formula of MAX is $M_{n+1}AX_n$, with A as an element from the group 13–14 in the periodic table [4]. It is known that MXenes consist of a hexagonal structure, which is similar to their parent MAX phases. MXenes have been explored for use in various fields such as energy storage devices, smart sensors, and electromagnetic shielding systems owing to their excellent hydrophilicity, electrical conductivity, fast ion transfer, and mechanical/chemical stability [5–8]. Carbon and transition metal atoms enable MXenes to exhibit superior pseudocapacitive behavior with good electrical conductivity [9]. In addition, the space between MX layers formed by removing A-layers in MAX phases can accommodate ions such as Li^+, which makes MXenes very promising for energy storage applications.

Supercapacitors include electrochemical double-layer capacitors (EDLCs) and pseudocapacitors, and display many attractive features including high power and fast charge/discharge rates, with long cyclic stability [10–12]. Carbon-based EDLCs store their charges electrostatically, which enables the fast and reversible adsorption/desorption of electrolyte ions on the highly porous carbon electrodes [13]. For pseudocapacitors, their energy storage mechanism relies on fast and reversible redox reactions near the surface of the electrode materials, such as conductive polymers or transition metal oxides. They exhibit greater energy density and lower power density compared to EDLCs, since faradaic redox reactions are intrinsically slower than electrosorption. Pseudocapacitance occurs through monolayer ion adsorption with charge transfer and redox reactions on the electrode surface [14]. Therefore, the surface coating of electroactive materials on the porous carbon materials can offer a good strategy to obtain pseudocapacitive performance effectively.

Electrospun carbon nanofibers (CNFs) are good candidates for electrode materials since they exhibit excellent surface porosity and electrical conductivity [15,16]. For synthesis, precursor polymers such as polyacrylonitrile (PAN) or aromatic polyimides are electrospun, activated under air and carbonized under inert gas. It is known that their surface properties and electrochemical performance can be determined by tailoring chemical structures of precursor polymers [17,18]. Electrospinning is known as a facile way to produce composite materials by mixing additives in the polymer solutions. It was reported that MXenes were added to polyvinyl alcohol (PVA) solution and then electrospun to produce MXene-embedded composite nanofibers to improve electrical properties [19]. MXene/CNF composite electrodes prepared by electrospinning and carbonizing MXene flakes in PAN solution were reported for use in supercapacitor electrodes [20–22].

For supercapacitor electrodes, charge storage takes place at the electrode/electrolyte interface, and thus surface properties of electrode materials are critical. Therefore, it is highly required to place redox-active materials such as MXenes on the electrode surface. For the previously reported MXene/CNF composite electrodes, MXenes were embedded inside the fibers, which may not contribute to redox reactions at the electrode/electrolyte interface [20–22]. The surface coating of redox-active materials on the carbon electrode has been achieved by electrodeposition [23,24], hydrothermal deposition [25], chemical grafting [26] or dip-coating [27]. Among them, dip-coating is a facile and cost-effective process to produce composite materials. It was reported that MXene-coated carbon microfibers were prepared via dip-coating for supercapacitor applications, and pre-treatment was conducted prior to dip-coating for better adhesion between MXene flakes and fibers. Dharmasiri et al. reported an electrochemical surface treatment to tailor the surface chemistry of carbon fibers to optimize the adhesion of MXene flakes [28]. In addition, a conductive binder was utilized for the facile adhesion of MXene flakes to carbon fiber surface [29].

This paper reports the preparation of MXene-coated CNFs by dip-coating electrospun CNFs in the MXene colloidal solution without pre-treatment. CNFs are chosen for their EDLC performance with high surface porosity and feasibility as binderless and conductive additive-free electrodes, while MXenes are chosen for their pseudocapacitive charge storage with abundant redox-active sites. Benefiting from these features, MXene-coated CNF electrodes manifest superior energy storage performance.

2. Experimental

For the synthesis of MXenes (Ti_3C_2), titanium aluminum carbide (MAX, Ti_3AlC_2) and dimethyl sulfoxide (DMSO) were purchased from SY innovation (Pyeongtaek, Republic of Korea) and Tokyo-Chemical Industry Co. (TCI, Tokyo, Japan), respectively. Hydrofluoric acid (HF, 48 wt. %) and ammonium hydroxide (NH_4OH) solution were purchased from Sigma-Aldrich Co. (St. Louis, MO, USA). For the synthesis of CNFs, polyacrylonitrile (PAN) and N,N-dimethylformamide (DMF) were obtained from Pfaltz & Bauer Inc. (Waterbury, CT, USA) and TCI, respectively. For electrochemical measurements, a 3-electrode set and sodium sulfate (>99.5%, Na_2SO_4) were purchased from Wizmac Co. (Daejeon, Republic of Korea) and TCI, respectively. All chemicals were used as received.

The synthesis of electrospun CNFs was carried out by electrospinning and thermal treatment (oxidative stabilization and the subsequent carbonization) of a precursor polymer,

PAN; 10 wt. % PAN was dissolved in DMF at 40 °C overnight, and then electrospun under a voltage range from 17 to 20 kV. The tip-to-collector distance was 15 cm and the speed of the drum collector was 100 rpm. Electrospun PAN nanofibers were then stabilized at 280 °C at a rate of 2 °C/min for 2 h under air flow and carbonized in 800 °C at a rate of 10 °C/min for 30 min under nitrogen.

To obtain MXene colloidal solution, 0.5 g MAX was added into 10 mL HF for 5 min and stirred for 24 h at room temperature. The solution was then centrifuged for 10 min at 8000 rpm. After centrifugation and washing 6 times, the precipitate was dispersed in 10 mL DMSO followed by sonication for 1 h. After centrifugation for 10 min at 8000 rpm and washing, the precipitate was dispersed in 50 mL distilled water and 5 drops of NH_4OH solution were added while stirring for 30 min. After centrifugation for 10 min at 8000 rpm, a dark supernatant was obtained. MXene-coated CNFs were obtained by dip-coating CNFs in the MXene colloidal solution at 40 °C for 5 h, then vacuum-dried at 60 °C for 12 h. This is represented as M-CNFs.

The surface morphology and elemental compositions of synthesized CNFs and M-CNFs were investigated by scanning electron microscopy (SEM) and energy dispersive X-ray spectroscopy (EDS), respectively, using an SU8220 (Hitachi, Chiyoda, Japan). The crystal structures of MXene and CNFs were determined by X-ray diffraction (XRD, MiniFlex, Rigaku, Japan) scanning from a 2θ range of 3 to 90 degrees at a rate of 5 degrees/min. Attenuated total reflection Fourier transform infrared (ATR-FTIR) spectroscopy was conducted to study their chemical structures on a VERTEX 80 V (Bruker, Bremen, Germany) using a germanium crystal substrate. Microstructures of M-CNFs were characterized with high-resolution transmission electron microscopy (HRTEM, Titan G2 ChemiSTEM Cs Probe, FEI Company, Hillsboro, OR, USA) after sonication in ethanol for 10 min. Surface properties were measured via nitrogen adsorption/desorption isotherms using a surface area analyzer (Belsorp-Max, MicrotracBEL, Osaka, Japan). The specific surface area was computed using the Brunaure–Emmett–Teller (BET) equation.

Supercapacitor performance was measured by assembling a 3-electrode set, which consisted of a Ag/AgCl reference electrode, a platinum counter electrode and 1M Na_2SO_4 aqueous solution as an electrolyte. As a working electrode, synthesized CNFs and M-CNFs were used without the use of any additives such as binders or conductive agents, which result in lower specific capacity. Electrochemical characterization was performed by obtaining cyclic voltammetry (CV) and galvanostatic charge–discharge (GCD) on a WBCS3000S electrochemical work station (Wonatech, Seoul, Republic of Korea). The specific capacitance (C_{sp}, F/g) was calculated by:

$$C_{sp} = \frac{I_d t_d}{mV} \tag{1}$$

where I_d is the discharge current (A), t_d is the discharge time (s), m is the mass of CNFs or M-CNFs (a working electrode (g)), and V is the working potential window (V). In addition, the cycling stability of CNF and M-CNF electrodes was studied by measuring capacitance retention via GCD testing for 5000 cycles. Energy (E, Wh/kg) and power densities (P, kW/kg) were also evaluated via GCD testing by applying Equations (2) and (3), respectively:

$$E = \frac{I_d t_d V}{2m} \tag{2}$$

$$P = \frac{E}{t_d} \tag{3}$$

3. Results and Discussion

Electrospinning and thermal treatment (pre-oxidation and the subsequent carbonization) of PAN dissolved in DMF was carried out to produce CNFs. An MXene/CNFs composite was prepared by dip-coating CNFs in the MXene (Ti_3C_2) colloidal solution synthesized by the HF etching of MAX (Ti_3AlC_2).

The microscopic structure and crystalline nature were investigated by XRD measurement, as shown in Figure 1. The XRD patterns of MAX and MXene are shown in Figure 1a. After HF etching, the strong peak of MAX at 2θ = 38.8° corresponding to the (104) plane obviously decreased, and a non-basal diffraction peak at 60.8° corresponding to the (110) plane appeared, confirming that a part of MAX reacted with HF to form Mxene [30]. It is also noticeable that the shift of the reflection peak (002) from 9.7° for MAX to 9.3° for MXenes indicates an increase in the d-spacing due to successful exfoliation by the etching of MAX phases in HF.

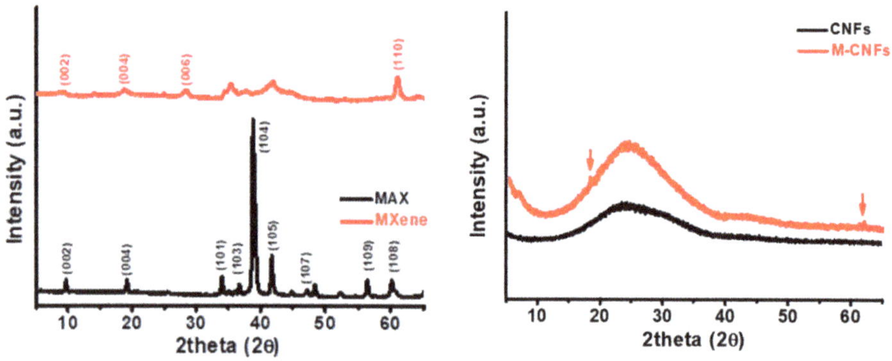

(a) (b)

Figure 1. XRD patterns of (**a**) MAX and MXene phases and (**b**) CNFs and M-CNFs.

Figure 1b displays the XRD patterns of CNFs and M-CNFs, and both exhibit a significant peak at 24.3°, indicating the formation of a graphite structure via the carbonization of PAN [31]. Different from CNFs, M-CNFs exhibit peaks at 18.4° and 61.7° (indicated by arrows) corresponding to the (004) and (110) plane of MXene, respectively, indicating the presence of MXenes in M-CNFs.

The chemical structures of CNFs and M-CNFs were characterized by FTIR analysis. As illustrated in Figure 2, the absorption peaks of C=C double bond and C–C single bond around 1550 cm^{-1} and 1130 cm^{-1} (indicated by black arrows) appeared in the spectra for both CNFs and M-CNFs. A peak at 535 cm^{-1} (indicated by a red arrow) corresponding to the bending vibration of Ti–O was seen for M-CNFs, while it was not found for CNFs. Indicative peaks of MXenes were clearly found in both the XRD pattern and FTIR spectrum of M-CNFs, which confirms that MXenes were well attached on the CNFs by the dip-coating method.

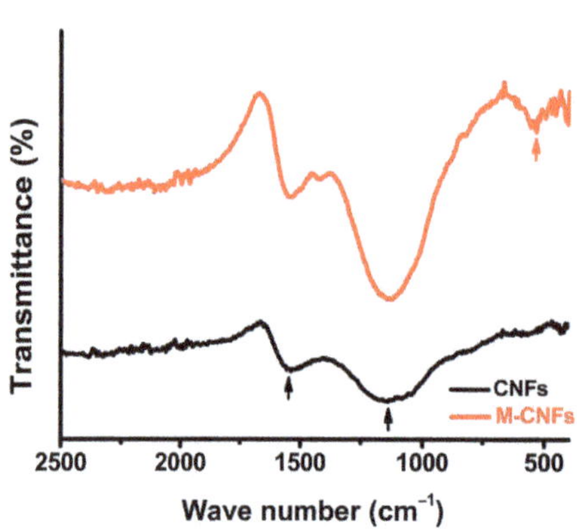

Figure 2. FTIR spectra of CNFs and M-CNFs.

The morphological structures of CNFs and M-CNFs were observed by SEM, as displayed in Figure 3. The successful fabrication of nanofibrous carbon structures through thermal treatment is confirmed by Figure 3a. Figure 3b shows the surface image of M-CNFs, and it is seen that MXenes were well placed on the CNF surface by the dip-coating of CNFs in the MXene colloidal solution. Figure 3c presents a low-magnification SEM image of M-CNFs, where MXene flakes are uniformly dispersed over the surface of dip-coated CNFs. Aqueous colloidal solutions of MXenes were used for dip-coating, since it is well-known that MXenes exhibit high dispersion stability in aqueous media due to their strong hydrophilicity [32,33]. It is also known that oxygen-containing functional groups exist on the CNF surface, resulting from the oxidative stabilization process [34]. As a result of the physisorption caused by these surface features of MXenes and CNFs, the stable dispersion of MXenes over the CNF surface was obtained. In Figure 3d, elementary analysis was carried out by using the EDS, which identified carbon and titanium as the elemental components in M-CNFs. This indicates that carbon is uniformly dispersed throughout the CNF mat, and titanium is clearly found in MXenes.

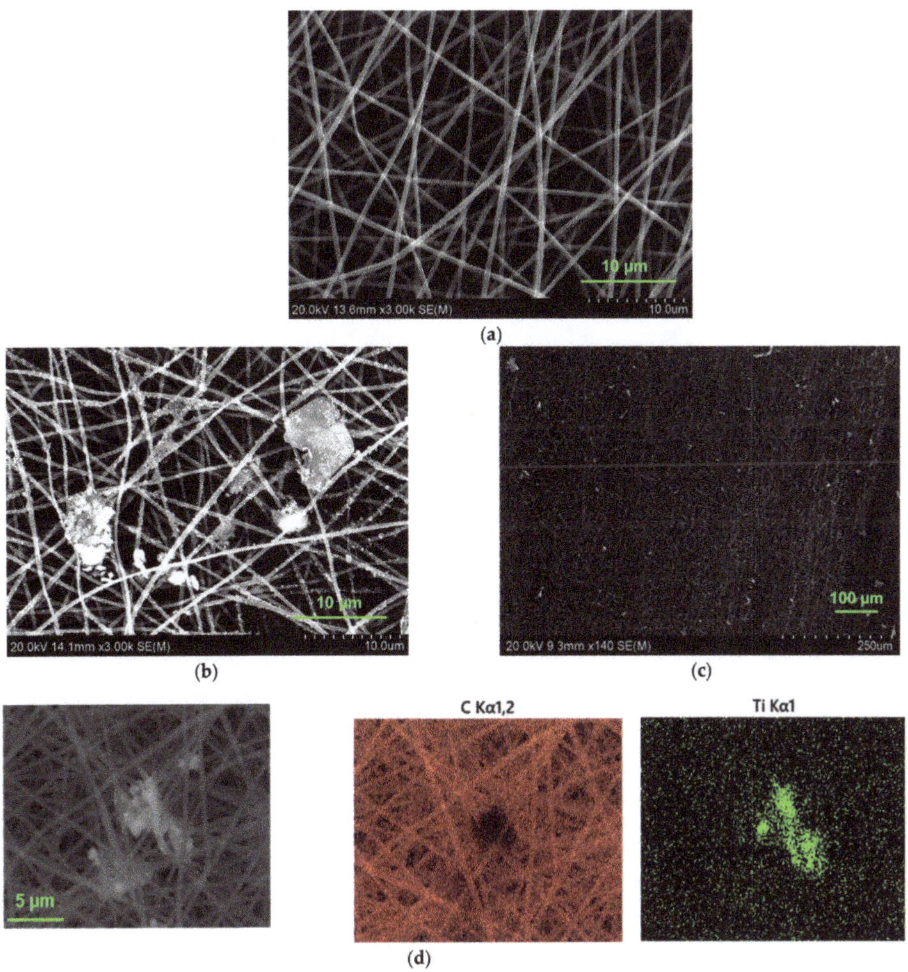

Figure 3. SEM images of (**a**) CNFs, (**b**) M-CNFs, (**c**) M-CNFs (low-magnification) and (**d**) EDS mapping images of M-CNFs.

To further observe microstructures, synthesized MXenes and M-CNFs were characterized by TEM, as displayed in Figure 4. A 2D nanosheet of the synthesized MXene is seen in Figure 4a, which confirms the successful exfoliation of MAX phases via HF etching. The higher-magnification image shows that the thickness of a single layer of MXene was 0.922 nm, which is consistent with the literature [35,36]. Figure 4b shows a TEM image of M-CNFs, and it is clearly seen that the MXene flake was attached to the surface of a single fiber.

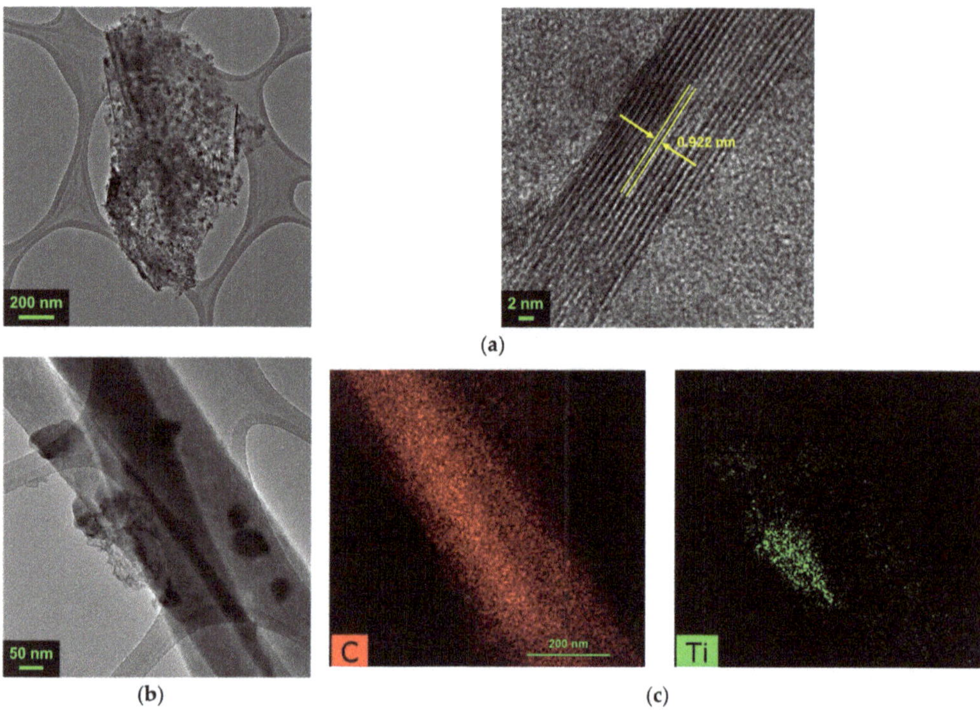

Figure 4. HR TEM images of (**a**) MXene and (**b**) M-CMFs, and (**c**) EDS mapping images of M-CNFs.

The surface properties of synthesized CNFs and M-CNFs were investigated by measuring the nitrogen adsorption and desorption isotherms, as presented in Figure 5. The isotherms for both CNFs and M-CNFs show a type I behavior, which is typically observed in microporous materials. The surface and pore characteristics are shown in Table 1. The CNFs synthesized via thermal treatment (oxidative stabilization and carbonization) of electrospun PAN showed a high specific surface area of 891.46 m^2/g, with a total pore volume of 0.415 cm^3/g. After dip-coating CNFs in the MXene colloidal solution, the surface area and pore volumes were decreased, since MXene flakes were placed on the surfaces of highly porous CNFs, which may have contributed surface-induced pseudocapacitance, and the pores of CNFs were covered by some MXene flakes larger than the fiber diameters of CNFs, as seen in Figure 3b,c. Still, M-CNFs displayed a high specific surface area of 663.03 m^2/g with a total pore volume of 0.283 cm^3/g. It was also found that the proportions of meso-pores in the pore diameter range of 2–50 nm for both CNFs and M-CNFs were around 10%.

The pseudocapacitive performance was investigated by assembling a three-electrode cell with Ag/AgCl and a Pt plate as the reference electrode and counter electrode, respectively, and with 1M Na_2SO_4 aqueous solution as an electrolyte. Synthesized CNFs and M-CNFs were applied as a binder and conductive agent-free working electrode. Figure 6 shows CV curves taken at various scanning rates ranging from 10 to 100 mV/s. The current densities of both CNFs and M-CNFs reduced as the scan rate was increased, since the interaction between the electrode surfaces and the electrolyte ions was limited by kinetic and mass transport parameters [37]. It is apparently demonstrated here that M-CNF electrodes exhibit larger current densities compared to CNFs in the whole range of applied voltages.

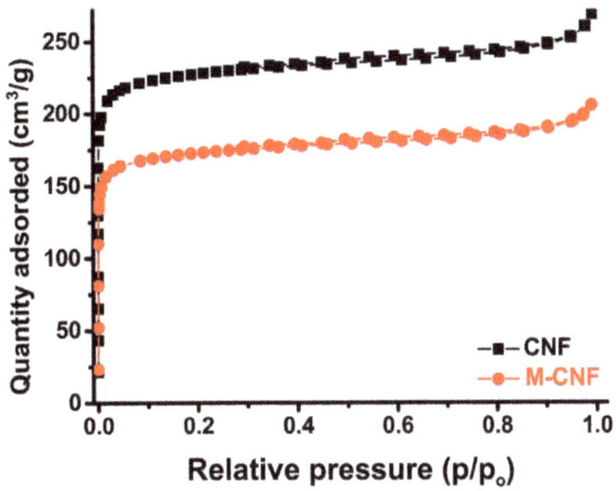

Figure 5. Nitrogen adsorption/desorption isotherms.

Table 1. Structural parameters of CNFs and M-CNFs.

	SSA [1] (m^2/g)	TPV [2] (cm^3/g)	V_{micro} [3] (cm^3/g)	V_{meso} [4] (cm^3/g)
CNFs	891.46	0.415	0.372	0.043
M-CNFs	663.03	0.317	0.283	0.034

[1] SSA: specific surface area. [2] TPV: total pore volume. [3] V_{micro}: micro-pore (<2 nm) volume. [4] V_{meso}: meso-pore (2–50 nm) volume.

Figure 6c displays CV curves of CNF and M-CNF electrodes at a scan rate of 10 mV/s for comparison. The CV curves of untreated CNFs are rectangular, without clear redox peaks, indicating typical EDLC characteristics due the potential-independent capacitance, while the CV curve for M-CNFs is quasi-rectangular with a shallow hump [38]. The higher capacitive behavior of M-CNFs, which is presented by the larger integrated area of CVs in Figure 6, originates from the pseudocapacitive charge storage induced by MXenes.

Energy performance was further investigated by carrying out galvanostatic charge–discharge testing, as seen in Figure 7a,b, under a current density from 0.5 to 5 A/g and potential range from 0 to 1 V. It is shown that M-CNF electrodes exhibited a significantly longer discharge time (1020 s) compared to that of the CNF electrodes (257 s) at 0.5 A/g. Figure 7c displays a Ragone plot, which presents energy density versus power density at different discharge rates. It an be seen that M-CNFs showed significantly higher energy and power densities at each discharge current density. Table 2 lists electrochemical properties, including maximum specific capacitance (C_{sp}) and energy and power densities (E and P, respectively), computed from the discharge curves in Figure 7. The specific capacitances of CNF and M-CNF electrodes are 130 and 514 F/g, respectively, at 0.5 A/g. It is also noted that the energy densities of CNF and M-CNF electrodes were 18.0 and 71.4 Wh/kg, respectively, at 0.5 A/g. Power density was estimated by an IR drop, which is the abrupt drop in potential at the beginning of discharge, and M-CNFs exhibited a higher value of 2.3 kW/kg at 5 A/g, compared to that of CNFs (1.7 kW/kg). This is attributed to the lower resistance of electrolytes and ion diffusion due to the presence of electrically conductive MXenes.

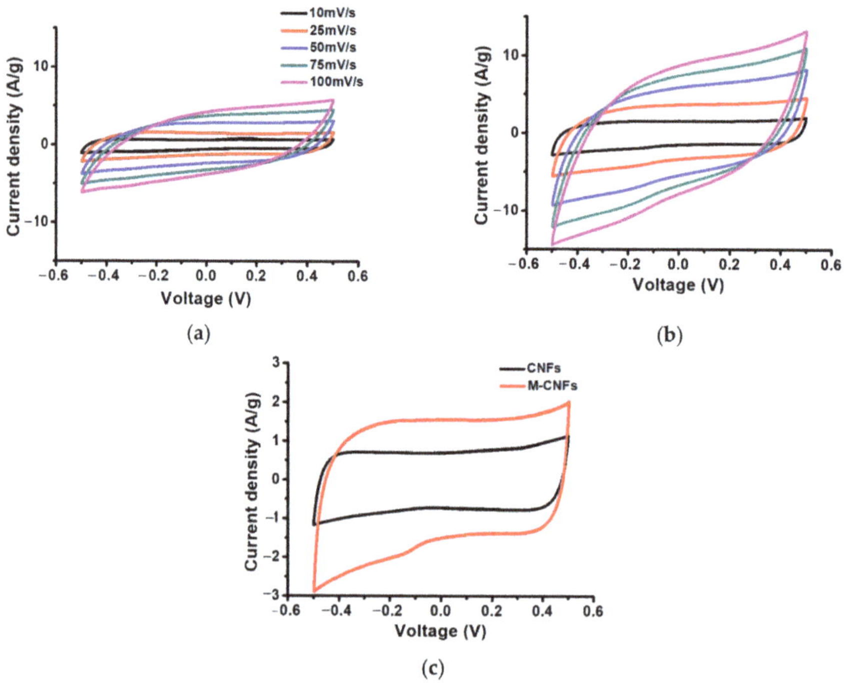

Figure 6. Cyclic voltammograms of (**a**) CNF, (**b**) M-CNF electrodes and (**c**) the comparison at a scan rate of 10 mV/s.

Table 2. Energy storage performance of CNFs and M-CNFs.

Sample	C_{sp} (F/g) @ 0.5 A/g	*P* (kW/Kg) @ 5 A/g	*E* (Wh/Kg) @ 0.5 A/g
CNFs	129.5	1.7	17.3
M-CNFs	514.0	2.3	70.1

The discharge curves of M-CNFs show inflection points, and this quasi-linear relationship between the potential and the time indicates pseudocapacitive behavior, which is consistent with CV curves [38]. The higher energy storage performance of M-CNFs is caused by the redox capacitance involving the electrochemical adsorption of ions at the surface of MXenes with a concomitant charge transfer [39].

The cyclic stability of CNF and M-CNF electrodes was investigated by measuring capacitance retention, and specific capacitance values under a current density of 1 A/g were collected for every 250 cycles by GCD testing, as displayed in Figure 8. Untreated CNFs showed slightly higher capacitance retention compared to M-CNFs. Due to the pseudocapacitive behavior of MXenes, M-CNFs showed greater capacitances and energy densities, originating from faradaic redox reactions at the interface between electrodes and electrolytes. This also resulted in lower cyclic stability. However, M-CNFs showed a capacitance retention of 90.7% after 5000 cycles, which is as high as that of untreated CNFs. The superior cyclic stability of both CNF and M-CNF electrodes can be confirmed by the high degree of reversibility. In addition, the electrochemical stability of M-CNF composite electrodes indicates the good attachment of MXenes on the CNF surface while GDC cycling test was performed for 5000 cycles.

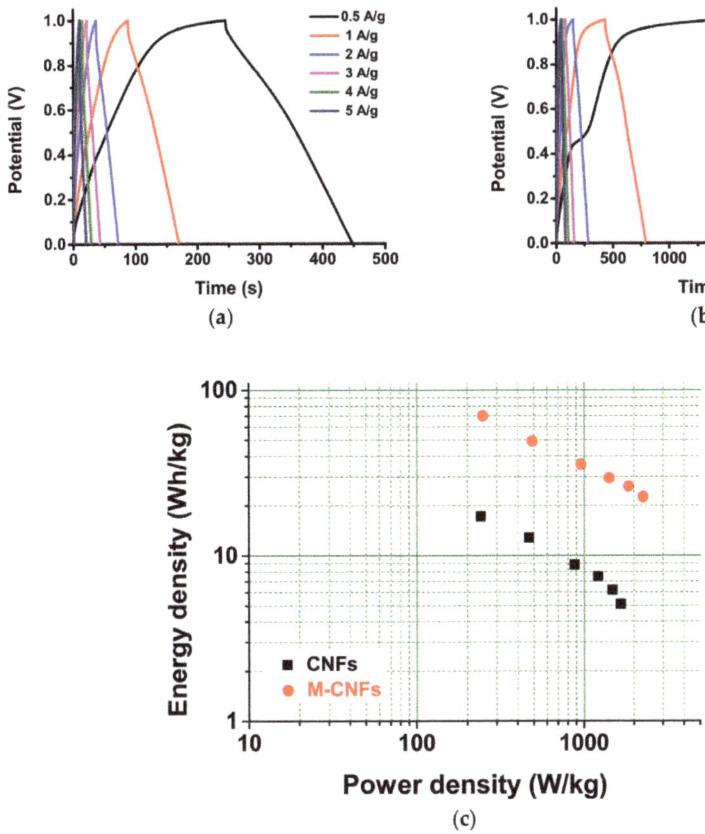

Figure 7. Charge–discharge curves of (**a**) CNF and (**b**) M-CNF electrodes, and (**c**) Ragone plot.

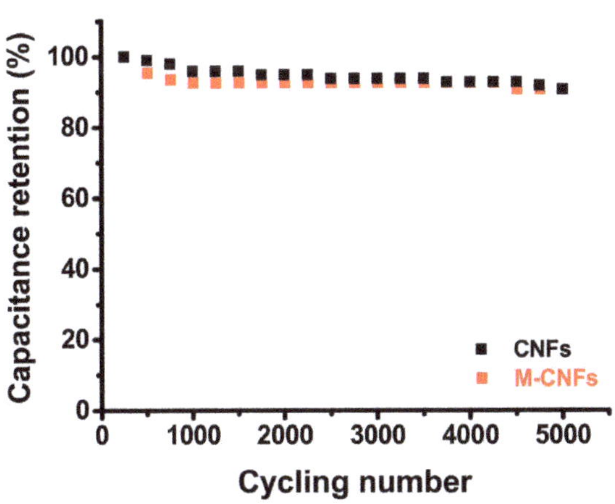

Figure 8. Cyclic stability of CNF and M-CNF electrodes.

MXene/carbon composites have been investigated since they exhibit superior electrochemical properties, owing to the pseudocapacitive behavior of MXenes and the high surface porosity of carbon. It is known that the high stacking of MXenes leads to prolonged ion transport pathways, resulting in limited energy storage performance, and carbon may effectively enlarge interlayer distances [40]. Li et al. reported the synthesis of graphene (rGO)/carbon nanotube (CNT) nanocomposites as supercapacitor electrodes to achieve compact structures with abundant ions access and fast charge transport [41]. These composite electrodes exhibited ultrahigh volumetric capacitances with high cyclic stability due to the surface redox reactions at the interlayer between the intercalated protons and the oxygen-containing functional groups. It was reported that MXene-bonded activated carbon was synthesized by encapsulating activated carbon at the interlayer of MXene, which resulted in enlarging the interlayer spacing between MXene layers and improving energy storage performance [42]. MXene/graphene composite films were electrostatically assembled using negatively charged MXene and positively charged graphene for ultrafast supercapacitors [43].

MXene-based composites using carbon fibers are a promising approach to developing flexible free-standing electrodes without the use of binders, which can cause poor adhesion, the agglomeration of active materials, excess mass of binders, and low conductivity. Wang et al. fabricated flexible yarn electrodes by biscrolling MXene with CNTs exhibiting superior specific capacitances, and energy and power densities [44]. Hu et al. reported the fabrication of MXene-coated carbon cloth by drop-casting for wearable smart electronics [45]. The M-CNFs used in this study also represent a great candidate for flexible free-standing electrodes showing superior energy storage performance.

Table 3 shows a comparison of the energy storage performances of the composite electrodes of MXenes and carbon in the literature. The specific capacitance of MXene-coated CNF electrodes in this work is higher than that of previously reported electrode materials.

Table 3. Comparison of energy storage performances of MXene/carbon composite electrodes in the literature.

Electrodes	Electrolyte	Scan Rate or Current Density	Specific Capacitance	Reference
MXene/rHGO	3M H_2SO_4	2 mV/s	438 F/g	[46]
MXene/rGO	3M H_2SO_4	2 mV/s	335.4 F/g	[43]
MXene/rGO	PVA/H_3PO_4	10 mV/s	327.5 F/g	[47]
MXene/CNTs	1M $MgSO_4$	2 mV/s	390 F/cm^3	[48]
MXene/CFs	1M H_2SO_4	5 mV/s	157 F/g	[28]
MXene/AC	1M Et_4NBF_4/AN	0.1 A/g	138 F/g	[42]
MXene/CNFs	1M H_2SO_4	50 mV/s	205 mF/cm^2	[20]
MXene/CNFs	1M H_2SO_4	2 mV/s	120 F/g	[21]
MXene/CNFs	3M KOH	1 A/g	249 F/g	[22]
M-CNFs	1M Na_2SO_4	0.5 A/g	514.0 F/g	This work

Composite electrodes of MXene and CNFs that were reported previously were prepared by adding MXene flakes in the precursor polymer solution prior to thermal treatment. Levitt et al. prepared MXene/CNF composite electrodes, of which the specific capacitance was 2.3 times higher than that of CNFs [20]. Hwang et al. also reported the preparation of MXene/CNF electrodes, and their specific capacitance was 3 times higher than that of pure CNFs [21]. The specific capacitance of M-CNFs in this work was 514.0 F/g, which is almost 4 times higher than that of untreated CNFs. This may be because highly conductive MXenes were placed on the surface of highly porous CNFs, which facilitated the access of electrolyte ions to the electrodes.

4. Conclusions

Supercapacitor performance is determined by the energy storage mechanism, which is mainly affected by applied electrode materials. To induce pseudocapacitive behavior driven by faradaic redox reactions at the interface between electrodes and the electrolyte, it is inevitable that the redox-active materials will be placed on the electrode surface. CNFs were chosen due to their high surface porosity with excellent EDLC performance, and the surfaces of CNFs were coated with MXene phases, which is a well-known redox-active component used to produce electrode materials with surface-induced pseudocapacitance.

In summary, the synthesis, characterization and electrochemical properties of MXene/CNF composites were conducted. They were synthesized by the dip-coating of CNFs prepared from the electrospinning and thermal treatment of PAN in the MXene colloidal solution, which was synthesized by the HF etching of MAX phase. The successful synthesis of an MXene/CNF composite was confirmed by investigating XRD and FTIR spectra, and SEM-EDS. Surface properties were observed via nitrogen adsorption/desorption, and M-CNFs displayed slightly lower surface area and pore volumes. Energy storage performance was measured using a three-electrode cell with Na_2SO_4 as an electrolyte, and M-CNF electrodes showed significantly superior supercapacitor performance, including specific capacitance (514 F/g) and energy and power densities (71.4 Wh/kg and 2.3 kW/kg, respectively), at 0.5 A/g, confirming them as promising electrode materials for supercapacitors.

Author Contributions: Methodology, S.K.K., S.A.K. and Y.S.H.; Validation, S.K.K. and S.A.K.; Formal analysis, S.K.K. and S.A.K.; Investigation, S.K.K.; Writing—original draft, K.-H.J.; Supervision, K.-H.J. All authors have read and agreed to the published version of the manuscript.

Funding: This work was financially supported by Daegu Catholic University (20211044).

Data Availability Statement: Data are contained within the article.

Conflicts of Interest: The authors declare no conflict of interest.

References

1. Kumbhakar, P.; Jayan, J.S.; Madhavikutty, A.S.; Sreeram, P.; Appukuttan, S.; Ito, T.; Tiwary, C.S. Prospective applications of two-dimensional materials beyond laboratory frontiers: A review. *IScience* **2023**, *26*, 106671. [CrossRef] [PubMed]
2. Naguib, M.; Kurtoglu, M.; Presser, V.; Lu, J.; Niu, J.; Heon, M.; Hultman, L.; Gogotsi, Y.; Barsoum, M.W. Two-dimensional nanocrystals produced by exfoliation of Ti_3AlC_2. *Adv. Mater.* **2011**, *23*, 4248–4253. [CrossRef]
3. Alhabeb, M.; Maleski, K.; Anasori, B.; Lelyukh, P.; Clark, L.; Sin, S.; Gogotsi, Y. Guidelines for synthesis and processing of two-dimensional titanium carbide ($Ti_3C_2T_x$ MXene). *Chem. Mater.* **2017**, *29*, 7633–7644. [CrossRef]
4. Barsoum, M.W. *MAX Phases: Properties of Machinable Ternary Carbides and Nitrides*; John Wiley & Sons: Hoboken, NJ, USA, 2013.
5. Champagne, A.; Charlier, J.-C. Physical properties of 2D MXenes: From a theoretical perspective. *J. Phys. Mater.* **2020**, *3*, 032006. [CrossRef]
6. Thakur, N.; Kumar, P.; Sati, D.C.; Neffati, R.; Sharma, P. Recent advances in two-dimensional MXenes for power and smart energy systems. *J. Energy Storge* **2022**, *50*, 104604. [CrossRef]
7. Ho, D.H.; Choi, Y.Y.; Jo, S.B.; Myoung, J.M.; Cho, J.H. Sensing with MXenes: Progress and prospects. *Adv. Mater.* **2021**, *33*, 2005846. [CrossRef] [PubMed]
8. Iqbal, A.; Sambyal, P.; Koo, C.M. 2D MXenes for electromagnetic shielding: A review. *Adv. Funct. Mater.* **2020**, *30*, 2000883. [CrossRef]
9. Xu, X.; Yang, L.; Zheng, W.; Zhang, H.; Wu, F.; Tian, Z.; Zhang, P.; Sun, Z. MXenes with applications in supercapacitors and secondary batteries: A comprehensive review. *Mater. Rep. Energy* **2022**, *2*, 100080. [CrossRef]
10. Trasatti, S.; Kurzweil, P. Electrochemical supercapacitors as versatile energy stores. *Platin. Met. Rev.* **1994**, *38*, 46–56. [CrossRef]
11. Halper, M.S.; Ellenbogen, J.C. *Supercapacitors: A Brief Overview*; The MITRE Corporation: McLean, VA, USA, 2006; Volume 1.
12. Yu, A.; Chabot, V.; Zhang, J. *Electrochemical Supercapacitors for Energy Storage and Delivery: Fundamentals and Applications*; Taylor & Francis: Abingdon, UK, 2013.
13. Pandolfo, A.G.; Hollenkamp, A.F. Carbon properties and their role in supercapacitors. *J. Power Sources* **2006**, *157*, 11–27. [CrossRef]
14. Wang, X.; Kajiyama, S.; Iinuma, H.; Hosono, E.; Oro, S.; Moriguchi, I.; Okubo, M.; Yamada, A. Pseudocapacitance of MXene nanosheets for high-power sodium-ion hybrid capacitors. *Nat. Commun.* **2015**, *6*, 6544. [CrossRef] [PubMed]
15. Kim, C. Electrochemical characterization of electrospun activated carbon nanofibres as an electrode in supercapacitors. *J. Power Sources* **2005**, *142*, 382–388. [CrossRef]
16. Zhang, B.; Kang, F.; Tarascon, J.-M.; Kim, J.-K. Recent advances in electrospun carbon nanofibers and their application in electrochemical energy storage. *Prog. Mater. Sci.* **2016**, *76*, 319–380. [CrossRef]
17. Kim, S.J.; Son, Y.J.; Jeon, B.; Han, Y.S.; Kim, Y.-J.; Jung, K.-H. Surface crosslinking of 6FDA-durene nanofibers for porous carbon nanofiber electrodes in electrochemical double layer capacitors. *Nanotechnology* **2020**, *31*, 215404. [CrossRef]
18. Lee, D.G.; Lee, B.C.; Jung, K.-H. Preparation of Porous Carbon Nanofiber Electrodes Derived from 6FDA-Durene/PVDF Blends and Their Electrochemical Properties. *Polymers* **2021**, *13*, 720. [CrossRef] [PubMed]

19. Sobolčiak, P.; Ali, A.; Hassan, M.K.; Helal, M.I.; Tanvir, A.; Popelka, A.; Al-Maadeed, M.A.; Krupa, I.; Mahmoud, K.A. 2D Ti3C2Tx (MXene)-reinforced polyvinyl alcohol (PVA) nanofibers with enhanced mechanical and electrical properties. *PLoS ONE* **2017**, *12*, e0183705. [CrossRef] [PubMed]
20. Levitt, A.S.; Alhabeb, M.; Hatter, C.B.; Sarycheva, A.; Dion, G.; Gogotsi, Y. Electrospun MXene/carbon nanofibers as supercapacitor electrodes. *J. Mater. Chem. A* **2019**, *7*, 269–277. [CrossRef]
21. Hwang, H.; Byun, S.; Yuk, S.; Kim, S.; Song, S.H.; Lee, D. High-rate electrospun Ti3C2Tx MXene/carbon nanofiber electrodes for flexible supercapacitors. *Appl. Surf. Sci.* **2021**, *556*, 149710. [CrossRef]
22. Yan, S.-x.; Luo, S.-h.; Wang, Q.; Zhang, Y.-h.; Liu, X. Rational design of hierarchically sulfide and MXene-reinforced porous carbon nanofibers as advanced electrode for high energy density flexible supercapacitors. *Compos. B Eng.* **2021**, *224*, 109246. [CrossRef]
23. Dirican, M.; Yanilmaz, M.; Fu, K.; Lu, Y.; Kizil, H.; Zhang, X. Carbon-enhanced electrodeposited SnO2/carbon nanofiber composites as anode for lithium-ion batteries. *J. Power Sources* **2014**, *264*, 240–247. [CrossRef]
24. Liu, Y.; Lu, Q.; Huang, Z.; Sun, S.; Yu, B.; Evariste, U.; Jiang, G.; Yao, J. Electrodeposition of Ni-Co-S nanosheet arrays on N-doped porous carbon nanofibers for flexible asymmetric supercapacitors. *J. Alloys Compd.* **2018**, *762*, 301–311. [CrossRef]
25. Wang, H.; Deng, J.; Chen, Y.; Xu, F.; Wei, Z.; Wang, Y. Hydrothermal synthesis of manganese oxide encapsulated multiporous carbon nanofibers for supercapacitors. *Nano Res.* **2016**, *9*, 2672–2680. [CrossRef]
26. Kotal, M.; Thakur, A.K.; Bhowmick, A.K. Polyaniline–carbon nanofiber composite by a chemical grafting approach and its supercapacitor application. *ACS Appl. Mater. Interfaces* **2013**, *5*, 8374–8386. [CrossRef] [PubMed]
27. Paleo, A.J.; Staiti, P.; Brigandì, A.; Ferreira, F.; Rocha, A.; Lufrano, F. Supercapacitors based on AC/MnO2 deposited onto dip-coated carbon nanofiber cotton fabric electrodes. *Energy Storage Mater.* **2018**, *12*, 204–215. [CrossRef]
28. Dharmasiri, B.; Usman, K.A.S.; Qin, S.A.; Razal, J.M.; Tran, N.T.; Coia, P.; Harte, T.; Henderson, L.C. Ti3C2Tx MXene coated carbon fibre electrodes for high performance structural supercapacitors. *Chem. Eng. J.* **2023**, *476*, 146739. [CrossRef]
29. Zhang, J.; Seyedin, S.; Gu, Z.; Yang, W.; Wang, X.; Razal, J.M. MXene: A potential candidate for yarn supercapacitors. *Nanoscale* **2017**, *9*, 18604–18608. [CrossRef] [PubMed]
30. Ghosh, A.; Pal, H.; Das, T.; Chatterjee, S.; Das, A. Synthesis and Characterization of MXene from MAX phase. *Mater. Today Proc.* **2022**, *58*, 714–716. [CrossRef]
31. Gupta, A.; Paliwal, D.; Bajaj, P. Acrylic precursors for carbon fibers. *J. Polym. Sci. Part. D Macromol. Rev.* **1991**, *31*, 1–89. [CrossRef]
32. Rozmysłowska, A.; Wojciechowski, T.; Ziemkowska, W.; Chlubny, L.; Olszyna, A.; Poźniak, S.; Tomkiewicz, K.; Jastrzębska, A. Colloidal properties and stability of 2D Ti3C2 and Ti2C MXenes in water. *Int. J. Electrochem. Sci.* **2018**, *13*, 10837–10847. [CrossRef]
33. Maleski, K.; Mochalin, V.N.; Gogotsi, Y. Dispersions of two-dimensional titanium carbide MXene in organic solvents. *Chem. Mater.* **2017**, *29*, 1632–1640. [CrossRef]
34. Wu, M.; Wang, Q.; Li, K.; Wu, Y.; Liu, H. Optimization of stabilization conditions for electrospun polyacrylonitrile nanofibers. *Polym. Degrad. Stab.* **2012**, *97*, 1511–1519. [CrossRef]
35. Yuen, A.C.Y.; Chen, T.B.Y.; Lin, B.; Yang, W.; Kabir, I.I.; Cordeiro, I.M.D.C.; Whitten, A.E.; Mata, J.; Yu, B.; Lu, H.-D. Study of structure morphology and layer thickness of Ti3C2 MXene with Small-Angle Neutron Scattering (SANS). *Compos. C Open Access* **2021**, *5*, 100155. [CrossRef]
36. Rahman, U.U.; Humayun, M.; Ghani, U.; Usman, M.; Ullah, H.; Khan, A.; El-Metwaly, N.M.; Khan, A. MXenes as emerging materials: Synthesis, properties, and applications. *Molecules* **2022**, *27*, 4909. [CrossRef]
37. Hashemi, M.; Rahmanifar, M.S.; El-Kady, M.F.; Noori, A.; Mousavi, M.F.; Kaner, R.B. The use of an electrocatalytic redox electrolyte for pushing the energy density boundary of a flexible polyaniline electrode to a new limit. *Nano Energy* **2018**, *44*, 489–498. [CrossRef]
38. Jiang, Y.; Liu, J. Definitions of pseudocapacitive materials: A brief review. *Energy Environ. Mater.* **2019**, *2*, 30–37. [CrossRef]
39. Kumar, S.; Rehman, M.A.; Lee, S.; Kim, M.; Hong, H.; Park, J.-Y.; Seo, Y. Supercapacitors based on Ti3C2Tx MXene extracted from supernatant and current collectors passivated by CVD-graphene. *Sci. Rep.* **2021**, *11*, 649. [CrossRef] [PubMed]
40. Zhu, Y.; Rajoua, K.; Le Vot, S.; Fontaine, O.; Simon, P.; Favier, F. Modifications of MXene layers for supercapacitors. *Nano Energy* **2020**, *73*, 104734. [CrossRef]
41. Li, K.; Wang, X.; Wang, X.; Liang, M.; Nicolosi, V.; Xu, Y.; Gogotsi, Y. All-pseudocapacitive asymmetric MXene-carbon-conducting polymer supercapacitors. *Nano Energy* **2020**, *75*, 104971. [CrossRef]
42. Yu, L.; Hu, L.; Anasori, B.; Liu, Y.-T.; Zhu, Q.; Zhang, P.; Gogotsi, Y.; Xu, B. MXene-bonded activated carbon as a flexible electrode for high-performance supercapacitors. *ACS Energy Lett.* **2018**, *3*, 1597–1603. [CrossRef]
43. Yan, J.; Ren, C.E.; Maleski, K.; Hatter, C.B.; Anasori, B.; Urbankowski, P.; Sarycheva, A.; Gogotsi, Y. Flexible MXene/graphene films for ultrafast supercapacitors with outstanding volumetric capacitance. *Adv. Funct. Mater.* **2017**, *27*, 1701264. [CrossRef]
44. Wang, Z.; Qin, S.; Seyedin, S.; Zhang, J.; Wang, J.; Levitt, A.; Li, N.; Haines, C.; Ovalle-Robles, R.; Lei, W. High-performance biscrolled MXene/carbon nanotube yarn supercapacitors. *Small* **2018**, *14*, 1802225. [CrossRef] [PubMed]
45. Hu, M.; Hu, T.; Cheng, R.; Yang, J.; Cui, C.; Zhang, C.; Wang, X. MXene-coated silk-derived carbon cloth toward flexible electrode for supercapacitor application. *J. Energy Chem.* **2018**, *27*, 161–166. [CrossRef]
46. Fan, Z.; Wang, Y.; Xie, Z.; Wang, D.; Yuan, Y.; Kang, H.; Su, B.; Cheng, Z.; Liu, Y. Modified MXene/holey graphene films for advanced supercapacitor electrodes with superior energy storage. *Adv. Sci.* **2018**, *5*, 1800750. [CrossRef] [PubMed]

47. Yang, Q.; Xu, Z.; Fang, B.; Huang, T.; Cai, S.; Chen, H.; Liu, Y.; Gopalsamy, K.; Gao, W.; Gao, C. MXene/graphene hybrid fibers for high performance flexible supercapacitors. *J. Mater. Chem. A* **2017**, *5*, 22113–22119. [CrossRef]
48. Zhao, M.-Q.; Ren, C.E.; Ling, Z.; Lukatskaya, M.R.; Zhang, C.; Van Aken, K.L.; Barsoum, M.W.; Gogotsi, Y. Flexible MXene/carbon nanotube composite paper with high volumetric capacitance. *Adv. Mater.* **2014**, *27*, 339–345. [CrossRef]

C Journal of Carbon Research

MDPI

Article

Insights into the Electrocatalytic Activity of Fe,N-Glucose/Carbon Nanotube Hybrids for the Oxygen Reduction Reaction

Rafael G. Morais [1,2,*], Natalia Rey-Raap [3], José L. Figueiredo [1,2] and Manuel F. R. Pereira [1,2,*]

1 LSRE-LCM—Laboratory of Separation and Reaction Engineering-Laboratory of Catalysis and Materials, Faculty of Engineering, University of Porto, Rua Dr. Roberto Frias, 4200-465 Porto, Portugal; jlfig@fe.up.pt
2 ALiCE—Associate Laboratory in Chemical Engineering, Faculty of Engineering, University of Porto, Rua Dr. Roberto Frias, 4200-465 Porto, Portugal
3 Instituo de Ciencia y Tecnología del Carbono, INCAR-CSIC, Francisco Pintado Fe 26, 33011 Oviedo, Spain; natalia.rey@incar.csic.es
* Correspondence: rgm@fe.up.pt (R.G.M.); fpereira@fe.up.pt (M.F.R.P.)

Abstract: Glucose-derived carbon hybrids were synthesized by hydrothermal treatment in the presence of oxidized carbon nanotubes. Additionally, iron and nitrogen functionalities were incorporated into the carbon structure using different methodologies. The introduction of iron and nitrogen in a single step under a H_2 atmosphere favored the formation of quaternary nitrogen and oxidized nitrogen, whereas the incorporation of nitrogen under an N_2 atmosphere after doping the hybrids with iron mainly produced pyridinic nitrogen. The samples were characterized by scanning electron microscopy, X-ray spectroscopy, adsorption isotherms, inductively coupled plasma optical emission spectrometry, and Raman spectroscopy. The presence of iron and nitrogen in the carbons increases the onset potential toward oxygen reduction in KOH 0.1 mol L^{-1} by 130 mV (0.83 V), in comparison to carbonized glucose, whereas the reaction mechanism shifts closer to a direct pathway and the formation of HO_2^- decreases to 25% (3.5 electrons). The reaction rate also increased in comparison to the carbonized glucose, as observed by the decrease in the Tafel slope value from 117 to 61 mV dec^{-1}. Furthermore, the incorporation of iron and nitrogen in a single step enhanced the short-term performance of the prepared electrocatalysts, which may also be due to the higher relative amount of quaternary nitrogen.

Keywords: carbon-based electrocatalyst; carbon nanotubes; oxygen reduction reaction; nitrogen-doping; iron

1. Introduction

Fuel cells represent one of the most promising alternatives to fossil fuels for sustainable green electrical energy conversion due to the low emission of pollutants, high efficiency, and non-polluting reaction products [1,2]. However, the oxygen reduction reaction (ORR) that takes place at the cathode presents highly sluggish kinetics and hinders the overall device operation [2–9]. Platinum-based electrocatalysts are the most commonly used to supply faster kinetics [3,5,6,8,10–14]. However, Pt is a scarce and expensive noble metal, contributing to the high costs associated with fuel cell production [15]. Consequently, the growing need to optimize green energy conversion devices has led to ever-increasing efforts to design non-noble metal electrocatalysts that display high electroactivity toward ORR and long-term durability.

The high versatility and lower price of carbon materials in comparison with metal-based materials make them one of the most interesting supports for electrocatalysts [2,16,17]. The modification of the physical and chemical properties of carbons can result in materials with higher electroactivity, which is essential for the ORR. In order to preserve the ecological friendliness of carbon materials, the combination of a green carbon source and an adequate synthesis method is essential. The use of biomass and hydrothermal carbonization (HTC) are presented as two interesting strategies for the abovementioned objective. Biomass is one of the most abundant carbon sources in the world and presents a relatively low cost. Furthermore, HTC is a low-cost methodology to obtain biomass-derived carbons using mild operating conditions via an environmentally friendly process [18–22]. This combination results in spherical particles that usually display low porosity and low electroactivity toward the ORR [19–22]. To overcome this drawback, tailoring the properties of carbons obtained by HTC is necessary. Carbonization and activation methods are usually employed to develop porosity, while the chemical properties can be modified by the incorporation of heteroatoms and/or transition metals. Nitrogen is considered the most effective heteroatom to enhance the carbon electrocatalyst's activity toward the ORR [19,21–26]. The presence of pyridinic nitrogen is important for O_2 bonding, whereas the incorporation of quaternary nitrogen favors oxygen dissociation [20]. Additionally, carbon nanotubes (CNTs) can also be added during the polymerization of glucose to increase the degree of polymerization and lower the surface's acidity and the graphitization level of the HTC carbons, in order to enhance the electron transfer kinetics during the ORR [19,22,27]. Nonetheless, the obtained electroactivities are still inferior to those of benchmark catalysts, and the incorporation of transition metals is essential to enhance the catalytic activity of these hybrids. Among transition metals, iron is largely reported to give rise to high performances toward the ORR [28,29], and the synergy between nitrogen and iron is one of the most important features in the reduction of oxygen and has been reported to achieve outstanding performances [30–34]. However, the effect of nitrogen and iron on carbon materials and their synergy largely depends on the structure and morphology of the supports [35] and the methodology employed for their incorporation [28]. Therefore, this work aims to shed light on the synergistic effect between nitrogen and iron incorporated by different methodologies and to evaluate which are the most promising for obtaining electrocatalysts with high efficiency and electroactivity toward the ORR. Accordingly, this work studies the effect of a wide range of iron concentrations added to carbonized glucose (CG). Additionally, the effect of the incorporation of a selected amount of iron on both CG and CG/CNT hybrids was studied, determining the influence of CNTs on the single incorporation of the transition metal. The effect of the methodology employed to introduce nitrogen to the CG and hybrids has also been assessed by following two different pathways to achieve Fe,N carbon samples: (i) a two-step method featuring the incorporation of iron oxide, followed by its reduction and the subsequent impregnation of melamine, as well as its decomposition into nitrogen functionalities by employing a thermal treatment under an inert atmosphere, or (ii) a one-step method featuring the impregnation of iron oxide and melamine followed by thermal treatment under a H_2 atmosphere. These methodologies allow us to understand how nitrogen functionalities are affected by the process and the base carbon employed (CG or hybrid), and, hence, the electroactivity toward the ORR.

2. Materials and Methods

2.1. Reagents

D(+)Glucose (HiMedia, Modautal, Germany, >99%), distilled water (produced on-site in a distillation unit), multiwalled CNT (Nanocyl, Sambreville, Belgium, C > 95%), and nitric acid (HNO_3, Honeywell, Charlotte, NC, USA ≥65%) were used to prepare the carbon materials. Fe_2O_3 (Sigma-Aldrich, St. Louis, MO, USA, ≥99%) and melamine ($C_3H_6N_6$, Sigma-Aldrich, 99%) were employed to functionalize the carbon supports. All gases used throughout this work were supplied by Air Liquide (Paris, France): nitrogen (N_2, 99.999%), hydrogen (H_2, 99.999%), and oxygen (O_2, 99.995%) were used during thermal treatments and/or electrochemical assessments. Nafion (Sigma-Aldrich, 5 wt.% in a mixture of water (15–20%) and aliphatic alcohols), absolute ethanol (VWR Chemicals, Radnor, PA, USA), and ultrapure water, produced by inverse osmose using Panice equipment, were used for the preparation of the electrocatalysts' dispersion, whereas methanol (VWR Chemicals, Radnor,

PA, USA, ≥99.9%) was employed during methanol tolerance measurements. Platinum on carbon black (Pt/C, Alfa-Aesar, Havervill, MA, USA, platinum nominal 20% on carbon black) was used as the ORR benchmark catalyst.

2.2. Preparation of Electrocatalysts

Glucose-derived carbon materials were prepared through the hydrothermal polymerization of glucose. Briefly, glucose (15.0 wt.%) was dispersed under sonication in distilled water (85.0 wt.%) for 30 min in a Teflon-lined reactor chamber. Afterward, the solution was closed in a Teflon-lined stainless-steel autoclave and underwent hydrothermal processing at 180 °C for 12 h. The resulting organic polymer was then abundantly washed with distilled water, filtered at room temperature, and dried at 100 °C for 16 h. The material was then carbonized under an N_2 atmosphere for 2 h at 800 °C using a heating ramp of 10 °C min^{-1}.

The chemical composition of the carbonized glucose sample was then modified by incorporating iron or iron/nitrogen according to three different methodologies, as detailed in Figure 1: (i) different amounts of Fe_2O_3 were impregnated into the carbonized material by incipient wetness impregnation and treated under an N_2 atmosphere for 2 h, then under a H_2 atmosphere for an additional 2 h at 650 °C, to achieve a certain final wt.% of iron on the carbon sample (0.5 wt.%, 1 wt.% or 5 wt.%, samples $CG_{0.5Fe}$, CG_{1Fe} and CG_{5Fe}, respectively); (ii) the prepared CG_{5Fe} sample was mixed with melamine in a 2:3 melamine/CG ratio, followed by thermal treatment at 650 °C in an N_2 atmosphere for 2 h (two-step method, sample N-CG_{5Fe}), and (iii) the simultaneous incorporation of 5 wt.% of iron oxide and melamine in a 2:3 melamine/CG ratio followed by a single thermal treatment, first in N_2 (2 h) and then under a H_2 atmosphere (2 h) at 650 °C (one-step method, sample $CG_{5Fe,N}$). All treatments were performed with a gas flow rate of 150 cm^3 min^{-1} and a heating rate of 10 °C min^{-1}.

Additionally, a series of hybrids were prepared by adding oxidized carbon nanotubes (CNT_O) during the hydrothermal polymerization of glucose, as described elsewhere [19]. A typical liquid-phase oxidation method was employed to obtain CNT_O. Briefly, a 7 mol L^{-1} solution of HNO_3 was used to oxidize CNTs for 3 h at the boiling temperature using a ratio of 75 mL of nitric acid per g of CNTs. The resulting oxidized material was then abundantly washed with distilled water until the water pH was attained and subsequently dried for 24 h at 80 °C. The hydrothermal hybrid was prepared following the same hydrothermal treatment procedure as that employed for the CG sample but using 14.7 wt.% of glucose, 0.3 wt.% of CNT_O, and 85.0 wt.% of distilled water as the initial dispersion. The resulting hydrothermal hybrid was carbonized under an inert atmosphere for 2 h at 800 °C and labeled CG/CNT_O. Afterward, the chemical composition of the material was also modified, as detailed in Figure 1. Fe_2O_3 was incorporated into CG/CNT_O by incipient wetness impregnation, followed by thermal treatment in N_2 (2 h) and then in H_2 atmosphere (2 h) at 650 °C to achieve a final iron content of 5 wt.% (CG/$CNT_{O,5Fe}$). Moreover, CG/$CNT_{O,5Fe}$ was mixed with melamine in a 2:3 melamine/CG/CNT_O ratio followed by the same thermal treatment as sample N-CG_{5Fe} (two-step method, N-CG/$CNT_{O,5Fe}$). Lastly, iron and nitrogen were simultaneously incorporated into CG/CNT_O following the same methodology as that used to prepare $CG_{5Fe,N}$ (one-step method, CG/$CNT_{O,5Fe,N}$).

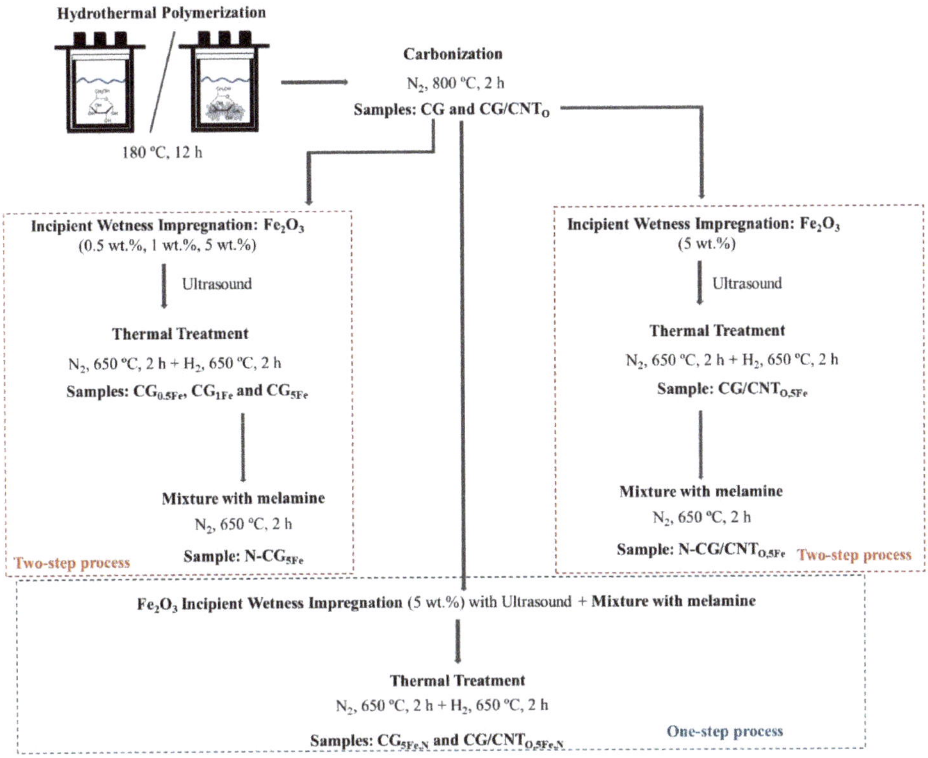

Figure 1. Scheme of the synthesis methodologies employed to prepare Fe,N carbon electrocatalysts.

2.3. Chemical and Textural Characterization

The contents of carbon, hydrogen, and nitrogen (C, H, and N) in the electrocatalysts were assessed by combusting the sample at 1050 °C using a Vario Micro Cube analyzer (Elementar GmbH, Langenselbold, Germany), whereas the oxygen content was measured by pyrolyzing the sample at 1450 °C in a Rapid Oxy Cube analyzer (Elementar GmbH, Langenselbold, Germany). To ensure the homogeneity and reproducibility of all results, triplicate analyses were performed for each material. To further understand the full composition of the prepared carbons, a Perkin Elmer Optima 4300 spectrometer (PerkinElmer, Waltham, MA, USA) was used to perform inductively coupled plasma optical emission spectrometry (ICP-OES) analyses to quantify the iron loading of Fe-CG and Fe-CG/CNT_O samples. This spectrometer was composed of an Echelle polychromator optic. To comprehend the chemical surface composition of each sample, a Kratos AXIS Ultra HAS spectrometer was used to perform X-ray photoelectron spectroscopy (XPS). The analyses were carried out in fixed analyzer transmission mode at 14 kV (175 W) using monochromatic radiation (Al Kα, 1486.7 eV). Regarding the textural properties, N_2 adsorption–desorption isotherms at -196 °C were obtained for each carbon material using an automated gas sorption analyzer (Autosorb iQ, Quantachrome, Boynton Beach, FL, USA). Before the analysis, all samples underwent degasification for 12 h at 150 °C. The specific surface area (S_{BET}) was calculated using the Brunauer–Emmett–Teller (BET) equation, whereas the volume of nitrogen adsorbed at a relative pressure of 0.99 (saturation point) was used to calculate the total pore volume (V_p). The microporous volume (V_{micro}) was determined by the *t*-plot method. All samples were analyzed by scanning electron microscopy (SEM) to observe their morphology, and the Fe,N hybrids were also analyzed by elemental map distribution. These analyses were performed in a Nova NanoSEM 200 (FEI Company, Hillsboro, OR,

USA). A conductive carbon double-sided adhesive tape was used to attach the powders of each sample onto an aluminum pin. During SEM analyses, an Everhart–Thornley detector (ETD) was used. Regarding elemental mapping, an Si(Li) detector was employed. Raman spectroscopy was performed using a monochromatic wavelength laser of 532 nm in an Alpha 300 apparatus (WiTec, Ulm, Germany), by directly measuring the spectra on the carbons.

2.4. Electrochemical Characterization

The synthesized electrocatalysts were electrochemically characterized using a PGSTAT 302 N potentiostat/galvanostat and a three-electrode beaker-type cell. The reference and counter electrodes used in this system were Ag/AgCl (KCl 3 mol L^{-1}) and a glassy carbon rod, respectively. A rotating ring-disk electrode (RRDE, Metrohm, Herisau, Switzerland, 24.9% collection efficiency, 5 mm in diameter) composed of a glassy carbon disk and a Pt ring was used as a working electrode for linear sweep voltammetry (LSV) and cyclic voltammetry (CV) assessments. Additionally, a rotating disk electrode (RDE, Metrohm, Herisau, Switzerland, 3 mm in diameter) also consisting of a glassy carbon disk was used as the working electrode to perform chronoamperometric measurements. These electrodes were prepared by drop-casting a suspension containing the carbon electrocatalysts to achieve a final mass loading of ca. 0.27 mg cm^{-2}. The suspension was prepared by mixing 4 mg of the electrocatalyst, 470 µL of ultrapure water (Millipore, Burlington, MA, USA), 80 µL of Nafion (5 wt% in lower aliphatic alcohols and water, Sigma-Aldrich, St. Louis, MO, USA), and 80 µL of ethanol (≥99.5%, VWR Chemicals, Radnor, PA, USA). Before drop casting, the suspension underwent sonication for at least 30 min until homogeneity was achieved.

The ORR activity was initially analyzed by LSV and CV measurements, which were performed in N_2-saturated 0.1 mol L^{-1} KOH followed by experiments in O_2-saturated electrolytes. The potentials applied during the electrochemical measurements were corrected to reversible hydrogen electrode (RHE) potentials by applying the following equation:

$$E(\mathrm{RHE}) = E\left(\frac{\mathrm{Ag}}{\mathrm{AgCl}}\right) + E^0\left(\frac{\mathrm{Ag}}{\mathrm{AgCl}}\right) + 0.059\mathrm{pH} = 0.964\mathrm{V}, \quad (1)$$

where E(RHE) is the potential corrected to the RHE, E(Ag/AgCl) is the potential applied to the reference electrode, and E^0(Ag/AgCl) is the potential difference between the Ag/AgCl (3 mol L^{-1}) electrode and the RHE (0.197 V).

An initial screening of redox reactions that could take place under the inert medium was accomplished by CV using a scan rate that ranged from 5 to 100 mV s^{-1} with no rotation. Afterward, LSVs were performed by rotating the working electrode from 400 to 3000 rpm and using a scan rate of 5 mV s^{-1}. Both measurements were performed from 1.15 V to −0.05 V (vs RHE). The O_2-saturated assessments were performed using the same procedure after completion of the experiments in the N_2-saturated electrolyte. The experimental current densities were obtained by subtracting the results obtained in N_2-saturated KOH from those measured in O_2-saturated electrolyte to guarantee that only the current densities of phenomena related to the reduction of oxygen were represented. From this corrected LSV, different crucial ORR parameters were calculated: (i) the onset potential (E_{onset}) at a current density of 0.1 mA cm^{-2} and (ii) the current density produced at 0.4 V vs RHE ($J_{0.4}$). The reaction mechanism was evaluated by using the current produced by the disk and ring of the RRDE, in which the ring potential was fixed at ca. 1.57 V vs RHE. These values can be used to calculate the formation of HO_2^- from a two-electron mechanism and the number of electrons exchanged (n_e) at each step of the reaction. Both were calculated with the currents obtained at 1600 rpm using Equations (2) and (3), respectively:

$$\mathrm{HO_2^-}(\%) = 200 \times \frac{\frac{I_R}{N}}{I_D + \frac{I_R}{N}}, \quad (2)$$

$$n_e = \frac{4I_D}{I_D + \frac{I_R}{N}}, \quad (3)$$

where I_R is the ring current (mA), I_D is the disk current (mA), and N is the collection efficiency (0.249).

Moreover, the short-term performance (10,000 s) of each material toward ORR was determined by applying 0.4 V vs RHE. Additionally, the methanol tolerance of the best-performing sample and the commercial Pt/C were tested by the addition of 1.5 mL of methanol ($\geq$99.9%, VWR Chemicals, Radnor, PA, USA) to the electrolyte during chronoamperometric measurements at ca. 1000 s. Moreover, samples CG and CG/CNT_O were also analyzed by electrochemical impedance spectroscopy (EIS) with an AC amplitude of 10 mV and a frequency range from 1 MHz to 10 mHz using a fully discharged cell at 0 V.

3. Results and Discussion

3.1. Chemical Properties

Elemental and ICP-OES analyses were performed to determine the bulk composition of the electrocatalyst. The results obtained are detailed in Table 1.

Table 1. Chemical composition determined by elemental analysis and ICP-OES.

Sample	C [wt.%]	H [wt.%]	O [wt.%]	N [wt.%]	Fe [wt.%]
CG	90.8	1.7	7.5	-	-
CG/CNT_O	92.7	0.8	6.5	-	-
$CG_{0.5Fe}$	93.4	1.3	5.0	-	0.3
CG_{1Fe}	92.2	1.0	5.4	-	1.4
CG_{5Fe}	87.4	1.3	6.2	-	5.1
CG/$CNT_{O,5Fe}$	90.6	0.8	3.7	-	4.9
N-CG_{5Fe}	81.9	1.3	8.1	3.3	5.4
N-CG/$CNT_{O,5Fe}$	83.4	1.1	6.4	5.1	4.0
$CG_{5Fe,N}$	82.2	1.4	8.6	2.0	5.8
CG/$CNT_{O,5Fe,N}$	85.4	1.0	7.2	2.2	4.2

The carbonized hydrothermal samples (CG and CG/CNT_O) were primarily composed of carbon (Table 1), with some hydrogen and oxygen. The oxygen content decreases in the presence of oxidized CNTs, indicating that these CNTs favored the polymerization of glucose during the hydrothermal treatment [19,27]. Consequently, the hybrid materials have a lower amount of functional groups than their glucose-derived counterparts, as previously observed by Fourier transform infrared analysis [19]. On the other hand, the incorporation of iron onto CG and CG/CNT_O leads to a decrease in the relative oxygen content due to thermal treatment applied after the wetness impregnation with iron oxide. Nonetheless, increasing the amount of iron precursor during the impregnation process leads to an increase in the oxygen content due to the oxygen contribution of the Fe_2O_3 added that is not totally reduced in the final material, especially employing CG as support. Furthermore, interesting differences are observed regarding the oxygen content due to the incorporation of nitrogen. The addition of melamine to CG_{5Fe} and CG/$CNT_{O,5Fe}$ leads to an increase in oxygen content from 6.2 wt.% and 3.7 wt.% to 8.6 wt.% ($CG_{5Fe,N}$) and 7.2 wt.% (CG/$CNT_{O,5Fe,N}$), respectively. Overall, the nitrogen functionalization of the carbon materials increases the oxygen content, whereas the methodology selected to incorporate Fe and N (a single-step or two-step process) does not seem to play an important role in the final oxygen content.

Regarding nitrogen content, the opposite effect is observed. In this case, the methodology employed to introduce N atoms to the carbons is crucial for the final relative content determined. The two-step methodology leads to a higher nitrogen bulk content of the carbon materials than a single-step process, which indicates that there are more possibilities for nitrogen bonding after the incorporation of iron due to the previous formation of stable species on the carbon matrix that will not compete with nitrogen upon the second thermal treatment. Lastly, the iron content of Fe,N CGs and Fe,N CG/CNT_O is very similar within the same carbon support type (5.4 vs. 5.8 wt.% and 4.0 vs 4.2 wt.%, respectively). Nonetheless, it is possible to observe that the use of CG led to a slightly higher incorporation of iron atoms than CG/CNT_O. Furthermore, interesting differences in the chemical composition were also detected due to the synthesis process. Regardless of the doping method employed, the addition of CNTs during the hydrothermal polymerization results in hybrids that present lower oxygen and iron contents but higher nitrogen percentages than their non-hybrid counterparts (Table 1), which suggests that the presence of CNTs favors the preferential incorporation of nitrogen functionalities instead of iron oxide.

To fully understand the differences due to the methodologies employed for the Fe,N incorporation process, the surface composition of these four samples was determined by XPS. The survey spectra are presented in Figure S1, and their chemical surface compositions are detailed in Table S1. The XPS surveys indicate the presence of carbon, oxygen and nitrogen, but do not indicate the presence of iron on the materials' surface analyzed, which suggests that iron is probably incorporated inside the porous structure of the carbons instead of on the surface [36], which is confirmed by the ICP-OES analysis (Table 1). The surface chemical composition obtained by XPS follows the same trend as that obtained by elemental analysis. The presence of CNTs or the use of the two-step method for incorporation results in a higher nitrogen content (Table S1). To further understand the nitrogen bonds formed during the thermal treatments, the high-resolution XPS N *1s* spectra of those samples prepared by the two-step method (N-CG_{5Fe} and N-CG/$CNT_{O,5Fe}$) and one-step method ($CG_{5Fe,N}$ and CG/$CNT_{O,5Fe,N}$) were deconvoluted into the main four nitrogen functionalities (Figure 2): pyridinic nitrogen (N-6) at 398.3 $\pm$ 0.1 eV, pyrrolic nitrogen (N-5) at 400.2 $\pm$ 0.1 eV, quaternary nitrogen (N-Q) at 401.6 $\pm$ 0.1 eV, and oxidized nitrogen (N-X) at 404.1 $\pm$ 0.2 eV [37].

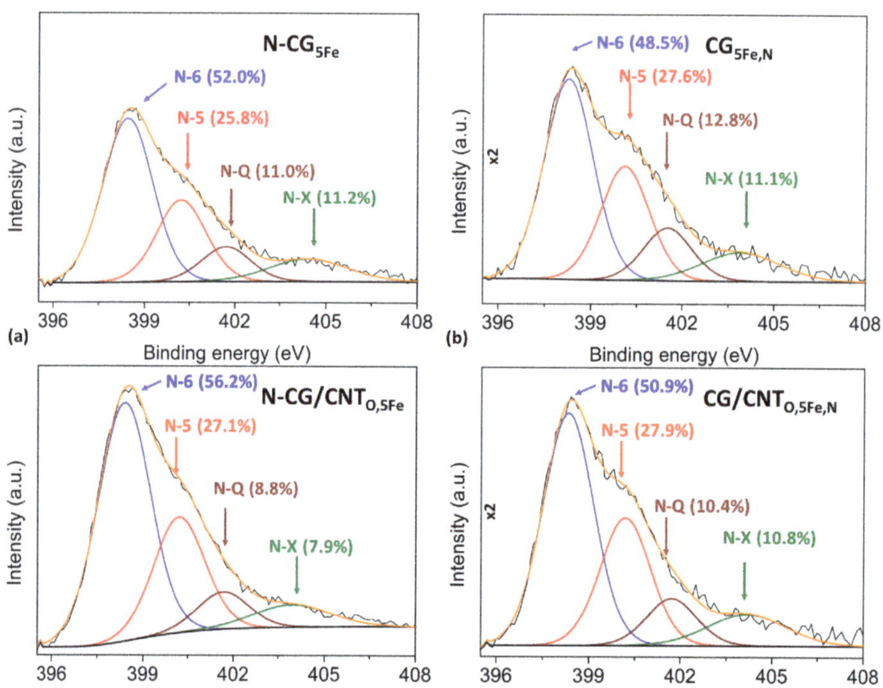

Figure 2. Deconvolution of the N *1s* X-ray photoelectron spectra of N-CG_{5Fe} (**a**), $CG_{5Fe,N}$ (**b**) N-CG/$CNT_{O,5Fe}$ (**c**) and CG/$CNT_{O,5Fe,N}$ (**d**) electrocatalysts.

Interesting differences are observed for the most common nitrogen functionalities. The relative composition of N-6 is higher in those materials in which the nitrogen precursor was treated under an inert atmosphere (N-CG_{5Fe} and N-CG/$CNT_{O,5Fe}$), while similar percentages of N-5 are observed in comparison with those samples treated under H_2 ($CG_{5Fe,N}$ and CG/$CNT_{O,5Fe,N}$). Furthermore, the percentage of N-Q increases after the treatment under H_2 (Figure 2b,d). The increase in N-Q and decrease in N-6 may be an effect of the time used during the one-step method (4 h) in comparison to the two-step method (2 h), which favors the transformation of N-6 into N-Q [38,39]. Moreover, all N-doped samples present N-X groups, which indicates that oxygen bound to nitrogen during the thermal treatments, as melamine does not present oxygen in its composition. This is in agreement with the elemental analysis results that show an increase in the oxygen content upon incorporating nitrogen functionalities.

Nitrogen adsorption–desorption isotherms performed at −196 °C were used to evaluate the textural properties of the prepared carbons and are displayed in Figure S2a,b, whereas the pore size distribution obtained from these isotherms is presented in Figure S2c,d. The main textural parameters calculated from the isotherms are shown in Table 2.

Table 2. Textural properties determined by N_2 adsorption/desorption isotherms at −196 °C.

Sample	S_{BET} (m^2 g^{-1})	V_p (cm^3 g^{-1})	V_{micro} (cm^3 g^{-1})
CG	468	0.20	0.18
CG/CNT_O	418	0.27	0.15
$CG_{0.5Fe}$	549	0.22	0.21
CG_{1Fe}	517	0.21	0.19
CG_{5Fe}	530	0.22	0.20
CG/$CNT_{O,5Fe}$	441	0.31	0.16
N-CG_{5Fe}	396	0.18	0.14
N-CG/$CNT_{O,5Fe}$	59	0.12	0.01
$CG_{5Fe,N}$	440	0.19	0.16
CG/$CNT_{O,5Fe,N}$	316	0.26	0.11

Sample CG exhibits a type I isotherm characteristic of microporous materials [40]. Nonetheless, the presence of the oxidized CNTs modifies the isotherm to a type IV [40,41], with a type H3 hysteresis loop characteristic of slit-shaped pores [41,42]. Moreover, the hybrid obtained (CG/CNT_O) presents a slight decrease in microporosity, which may have been caused by the absence of microporosity in the CNTs. The incorporation of a 5 wt.% of iron led to an increase in the S_{BET} for both CG (from 468 to 530 m^2 g^{-1}) and CG/CNT_O (from 418 to 441 m^2 g^{-1}), as well as in the pore volume (from 0.20 to 0.22 cm^3 g^{-1} and 0.27 to 0.31 cm^3 g^{-1}, respectively), which suggests that iron may be acting as an activating agent [43]. Regarding the pore size distribution, all of these samples are mainly composed of micropores (<2 nm) and some mesopores, which is in agreement with the hysteresis loop observed. The hybrid displays a slight shift of its pore widths to higher values than the CG sample (1.6 nm vs 1.5 nm, respectively) and a lower micropore volume (Figure S2c). The addition of iron leads to a slight decrease in the pore width, leading to micropores lower than 1 nm, and an overall increase in micropore volume for CG and hybrid samples. Regardless of the methodology employed to dope the carbons, Fe,N incorporation decreases the microporosity, especially in those samples prepared by the two-step method. This effect may be due to the time employed for the thermal treatments. In the two-step process, the melamine was only treated for 2 h under an N_2 atmosphere, while in the one-step process, the treatment took place for 2 h under N_2 and 2 h under H_2. This increase in residence time probably led to a higher degree of decomposition of the nitrogen precursor. The N-containing surface groups may have blocked some of the pores, thereby hindering the access of N_2 to the innermost pores during the isotherm measurements [20,36]. This is more evident for sample N-CG/$CNT_{O,5Fe}$ which displayed the highest S_{BET} and microporosity loss of all N-doped samples. This may be related to the higher nitrogen content of this sample in comparison to the other N-doped carbons synthesized, which increased the likelihood of having its pores blocked. This is also displayed in Figure S2d, in which sample N-CG/CNT$_{O,5Fe}$ presents a very low micropore contribution ranging from 1.0 to 1.7 nm pore width. Moreover, the addition of melamine after the incorporation of iron (N-CG$_{5Fe}$) leads to a wider pore size distribution than the simultaneous incorporation of nitrogen and iron (CG$_{5Fe,N}$), which could indicate that some morphological differences may exist between these two processes.

The morphological changes that occur during both the hydrothermal processes and the subsequent incorporation of iron (5 wt.%) and nitrogen were observed by scanning electron microscopy (SEM). The corresponding micrographs are displayed in Figure 3.

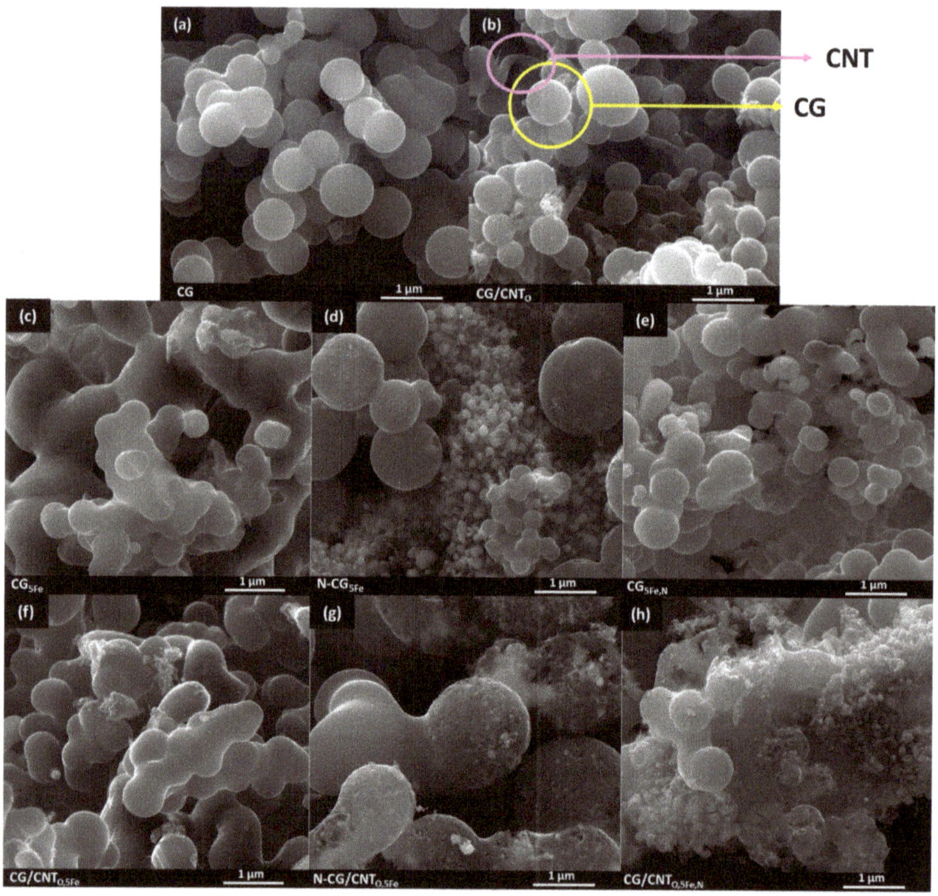

Figure 3. SEM micrographs of CG (**a**), CG/CNT$_O$ (**b**), CG$_{5Fe}$ (**c**), N-CG$_{5Fe}$ (**d**), CG$_{5Fe,N}$ (**e**), CG/CNT$_{O,5Fe}$ (**f**), N-CG/CNT$_{O,5Fe}$ (**g**) and CG/CNT$_{O,5Fe,N}$ (**h**) electrocatalysts.

Sample CG is composed of spheres without defects (Figure 3a), which is a typical morphology of hydrothermal carbons [44,45]. The incorporation of CNT$_O$ during hydrothermal polymerization leads to a combination of the carbon spheres and the CNTs, resulting in hybrid materials (Figure 3b). The addition of iron to the pristine samples (CG and CG/CNT$_O$) results in materials with interesting differences in their morphology. The micrographs of CG$_{5Fe}$ (Figure 3c) and CG/CNT$_{O,5Fe}$ (Figure 3f) exhibit some of the spheres observed in the non-doped samples but also a large number of carbon spheres that have agglomerated into a single large cluster. The incorporation of nitrogen results in materials with different morphologies (Figure 3d,e,g,h) compared to those of the monometallic samples (Figure 3c,f), with the appearance of several defects on the surface of the spheres (Figure 3d,e,g,h), probably due to the formation of C-N bonds during the decomposition of the melamine. Moreover, sample N-CG_{5Fe} exhibits spheres with different sizes, which may contribute to a wider pore size distribution than the more homogeneous $CG_{5Fe,N}$. Additionally, the iron dispersion on the carbon materials, specifically in Fe,N hybrids (N-$CG/CNT_{O,5Fe}$ and $CG/CNT_{O,5Fe,N}$), was analyzed by elemental mapping (Figure S3). The resulting micrographs indicate that iron is well dispersed in the area investigated, which indicates the successful incorporation of the metallic element across the carbon support.

To further address the changes attributed to the presence of CNT_O and/or the incorporation of iron and/or nitrogen, Raman spectroscopy was performed. The resulting spectra are exhibited in Figure S4. All carbons exhibit two main peaks: one associated with the carbon structure's basal plane defects (D band) at ca. 1350 cm^{-1} and another peak attributed to the material's graphitic domains (G band) at ca. 1580 cm^{-1}. These two main bands were further deconvoluted to gain some insight into the number of defects of each material. The integration of the peak areas resulting from the deconvolution (A_D and A_G) was then used to calculate the A_D/A_G ratio (Table S2). As expected, the presence of CNT_O decreases the value of the A_D/A_G ratio in comparison to CG, due to the highly graphitic nature of the CNTs. The incorporation of iron does not significantly modify the structure of CG. However, the incorporation of 5 wt.% of iron into the CG/CNT_O hybrid results in a higher A_D/A_G ratio, which may be attributed to the anchoring of the precursor nanoparticles on the carbon structure, which leads to a more disordered matrix [46]. The incorporation of nitrogen by the two-step method (N-CG_{5Fe} and N-$CG/CNT_{O,5Fe}$) produces different variations depending on the pristine carbon (CG or CG/CNT_O). N-CG_{5Fe} presents a slightly lower A_D/A_G ratio (2.36) than CG (2.44) and CG_{5Fe} (2.41), which indicates that each step of the electrocatalyst preparation contributes to increasing the degree of order. Contrarily, N-$CG/CNT_{O,5Fe}$ presents a higher A_D/A_G ratio (2.93) than both CG/CNT_O (2.22) and $CG/CNT_{O,5Fe}$ (2.57), which indicates that the precursor added is generating defects in the structure [19]. On the other hand, the simultaneous incorporation of iron and nitrogen ($CG_{5Fe,N}$ and $CG/CNT_{O,5Fe,N}$) results in a slight increase in the A_D/A_G ratio in comparison to their undoped counterparts (CG and CG/CNT_O, respectively) which indicates that the number of defects of the catalysts is increased during the incorporation process. These results indicate that the carbon support, precursors, and thermal treatments applied to synthesize these electrocatalysts are important parameters in the formation of defects.

3.2. *Electrochemical Assessments*

Oxygen Reduction Reaction

The synthesized electrocatalysts were initially screened toward the ORR by CV in a rotating ring-disk electrode (RRDE), using a scan rate range between 5 and 100 mV s^{-1} in inert and O_2-saturated electrolyte without rotation. Measurements under an N_2-saturated medium (Figure S5) displayed featureless cyclic voltammograms, which indicate that no electrocatalytic activity took place under these conditions. However, CVs performed in O_2-saturated KOH (Figure S5) show well-defined cathodic peaks, which confirms the electroactivity of all electrocatalysts toward ORR. Nonetheless, electrochemical differences are observed between samples due to their physico-chemical properties; hence, the performance of all catalysts was assessed by LSV to correlate the electrochemical results and the properties of the materials. The LSVs of the electrocatalysts and Pt/C recorded in the O_2-saturated 0.1 mol L^{-1} KOH electrolyte at 1600 rpm, and the respective number of electrons exchanged are plotted in Figure 4a,b and Figure 4c,d, respectively. The main electrochemical parameters obtained from the LSVs are exhibited in Table 3.

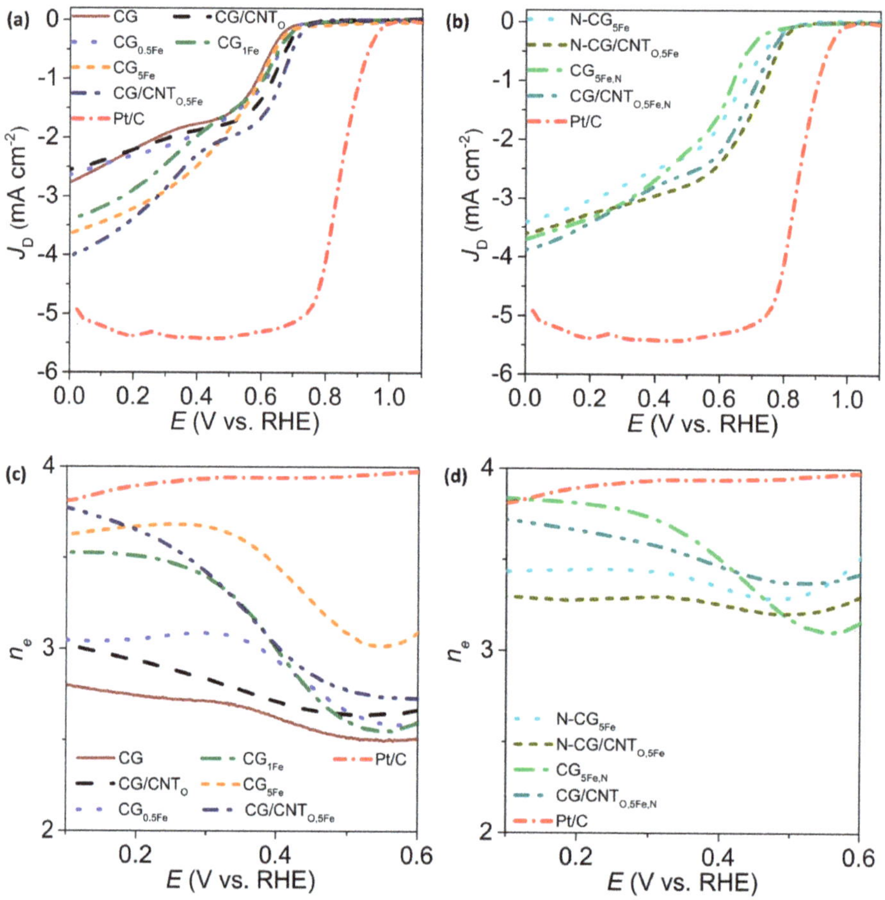

Figure 4. Disk current density produced by undoped and monometallic (**a**) and Fe,N (**b**) electrocatalysts at 1600 rpm in O_2-saturated electrolyte (0.1 mol L^{-1} KOH) and number of electrons exchanged during the oxygen reduction reaction for undoped and monometallic (**c**) and Fe,N (**d**) electrocatalysts.

Table 3. Main electrochemical parameters: onset potential (E_{onset}), current density ($J_{0.4}$), hydrogen peroxide (HO_2^-) formation, the number of electrons exchanged (n_e) and Tafel slope values for the synthesized materials and commercial Pt/C.

Sample	E_{onset} (V)	$J_{0.4}$ (mA cm^{-2}) [a]	HO_2^- (%) [a]	n_e [a]	Tafel Slope (mV dec^{-1})
CG	0.70	1.77	69	2.6	117
CG/CNT_O	0.73	1.87	64	2.7	77
$CG_{0.5Fe}$	0.71	1.92	53	2.9	97
CG_{1Fe}	0.72	1.97	50	3.0	67
CG_{5Fe}	0.75	2.47	27	3.5	64
CG/$CNT_{O,5Fe}$	0.74	2.29	49	3.0	45
N-CG_{5Fe}	0.81	2.50	33	3.3	52
N-CG/$CNT_{O,5Fe}$	0.83	2.95	37	3.3	63
$CG_{5Fe,N}$	0.79	2.67	25	3.5	62
CG/$CNT_{O,5Fe,N}$	0.81	2.79	27	3.5	61
Pt/C	0.98	5.41	3	3.9	62

[a]—Calculated at 0.4 V vs RHE.

The base CG and the hybrid CG/CNT_O were tested to evaluate their performance toward the ORR. The presence of CNT_O in the carbon structure shifts the E_{onset} by 30 mV from 0.70 V to 0.73 V, probably due to the higher number of active sites that were formed during the synthesis process. Furthermore, this hybrid exhibits an increase in $J_{0.4}$ of 6% and a small shift of the reaction mechanism from the two- to the four-electron pathway. These improvements are probably due to an increase in the electrical conductivity of the electrocatalyst [19]. To further investigate this hypothesis, electrochemical impedance spectroscopy (EIS) was performed on samples CG and CG/CNT_O at 0 V. The resultant Nyquist plots are displayed in Figure S6. These plots demonstrate that the presence of CNT_O led to a smaller semicircle, which indicates a decrease in the charge transfer resistance. This analysis suggests that the glucose/CNT hybrid has a higher electrical conductivity than the CG sample, which corroborates the previous statement. Furthermore, a decrease in the Tafel slope value from 117 mV dec^{-1} (CG) to 77 mV dec^{-1} (CG/CNT_O) is observed, which indicates a faster reaction rate [47].

The incorporation of different amounts of iron into the CG sample leads to an increase in E_{onset} and $J_{0.4}$. In fact, a direct relationship between these parameters and the amount of iron present in the electrocatalysts has been found (Figure S7). Furthermore, a decrease in the formation of HO_2^- and an increase in the number of electrons exchanged is also observed by increasing the percentage of iron. Similarly, the addition of 5 wt% of iron to the hybrid (CG/$CNT_{O,5Fe}$) increases the onset potential, current density, and number of electrons exchanged and decreases the formation of HO_2^-. The incorporation of iron also decreases the Tafel slope value with increasing iron content from 97 mV dec^{-1} ($CG_{0.5Fe}$) to 64 mV dec^{-1} (CG_{5Fe}), which is comparable to the benchmark Pt/C at low overpotentials [47,48].

Regarding the incorporation of nitrogen, several interesting differences are observed in the ORR, which are probably correlated to the amount and type of functionalities present in each material. The incorporation of nitrogen increases the onset potential from 0.75 V (CG_{5Fe}) to 0.81 V (N-CG_{5Fe}) and 0.79 V ($CG_{5Fe,N}$), and from 0.74 V (CG/$CNT_{O,5Fe}$) to 0.83 V (N-CG/$CNT_{O,5Fe}$) and 0.81 V (CG/$CNT_{O,5Fe,N}$). The incorporation of nitrogen after iron (two-step method) results in a higher increase in the onset potential than the single-step method. This difference may be explained by the presence of larger amounts of N-6 in samples N-CG_{5Fe} and N-CG/$CNT_{O,5Fe}$ than $CG_{5Fe,N}$ and CG/$CNT_{O,5Fe,N}$, respectively, which is believed to govern the E_{onset} [49]. Regarding the current density of these four samples, very similar values are observed for all samples across the potentials applied. The very small differences observed in the LSVs (Figure 4b) may be due to the slightly different reaction mechanisms which gradually shift the LSV form from a two-plateau (two-electron mechanism) to a single plateau (four-electron mechanism) [20]. The use of a single-step functionalization leads to fewer side products and a larger number of electrons transferred than a two-step functionalization. This may be due to the higher amount of nitrogen incorporated by the latter which may be acting as an ORR active site and competing with iron, leading to a small shift in the reaction mechanism. Once again, the incorporation of CNT_O improves the E_{onset} but does not have any significant effect on the reaction mechanism, as N-CG/$CNT_{O,5Fe}$ and CG/$CNT_{O,5Fe,N}$ present the same values of HO_2^- and number of electrons transferred as their counterparts. Overall, N-6 functionalities increase the E_{onset}, whereas the number of N-Q groups favors a more direct reaction mechanism (four electrons) and decreases the formation of HO_2^-. In comparison to similar N-doped carbons prepared using glucose as carbon source and melamine as nitrogen precursor [19], these materials (N-CG/$CNT_{O,5Fe}$ and CG/$CNT_{O,5Fe,N}$) display a higher E_{onset} (0.83 V vs. 0.79 V vs. RHE), higher current density (3.0 mA cm^{-2} vs. 2.5 mA cm^{-2}), and a lower formation of HO_2^-. All Fe,N carbon samples present a Tafel slope value close to 60 mV dec^{-1}, which indicates that the synthesized samples present a similar reaction rate to that of the Pt/C catalyst. Table S3 presents the ORR parameters of glucose-derived carbons published in the literature. Although the electrocatalytic activities obtained in this work are slightly inferior to those previously reported, it should be noted that the methodologies employed and reagents used are more environmentally friendly than most of the presented literature results, without requiring the use of templates or complex procedures. Moreover, the reaction rate observed from the Tafel slope values demonstrates that the glucose/CNT hybrids display very similar kinetics to those of the benchmark catalysts and other glucose-derived electrocatalysts reported in the literature.

To further evaluate the performance of the electrocatalysts toward the ORR, another critical parameter was analyzed, namely the short-term stability. Therefore, chronoamperometry measurements were performed on all electrocatalysts by applying 0.4 V (vs RHE) for 10,000 s. The results obtained are exhibited in Figure 5.

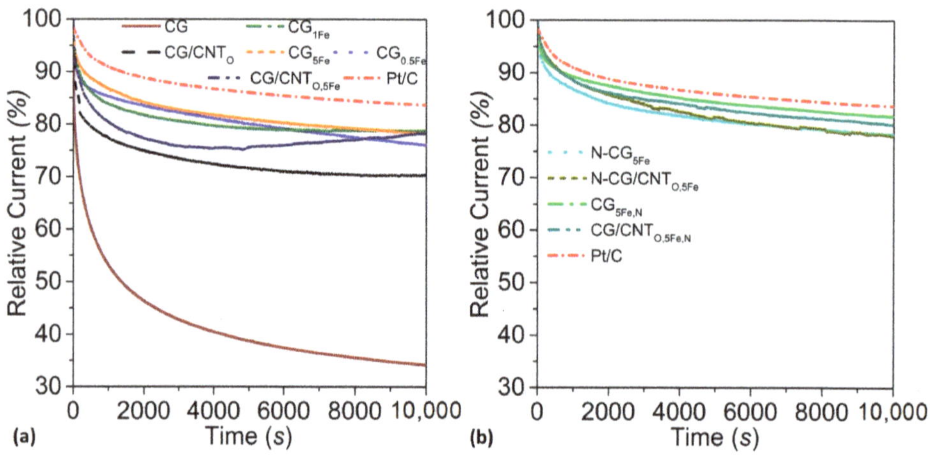

Figure 5. Chronoamperometry measurements (0.4 V for 10,000 s) for undoped and monometallic electrocatalysts (**a**), and Fe,N electrocatalysts (**b**) in an O_2-saturated 0.1 mol L^{-1} KOH electrolyte at 1600 rpm.

CG exhibits very low short-term stability, losing 66% of its current density after 10,000 s, which is not ideal for upscale applications. The presence of CNT_O increases the current density retention of the electrocatalysts by 36%, which can be attributed to the higher performance stability of the CNTs [19] and the modification of the carbon properties during the synthesis. Nonetheless, this value is still not ideal for long-term applications. It should be noted that there is a large amount of HO_2^- formed with both electrocatalysts (Table 3), which may lead to the oxidation of the carbon surface [50–52], hindering their ability for O_2 adsorption and, consequently, their performance toward the ORR. The incorporation of iron leads to an increase in the catalysts' stability, as iron species act as the main active centers toward the ORR and highlight a lower degradation level in comparison to the undoped carbons. Furthermore, interesting differences are observed due to the methodology followed to incorporate iron and nitrogen. Samples prepared by incorporating iron and then nitrogen (two-step method) display the same stability as their monometallic counterparts. This indicates that even though different active centers are present on the carbon (nitrogen and iron), the overall degradation of these sites maintains the same level as the samples without nitrogen. On the other hand, samples prepared by the simultaneous incorporation of iron and nitrogen functionalities have higher stability, which may be due to the higher relative content of N-Q. These functional groups present strong bonds that may favor the retention of the current density and thus the better short-term application of these catalysts. Overall, the Fe,N samples using the single-step methodology ($CG_{5Fe,N}$, and $CG/CNT_{O,5Fe,N}$) display very similar current density retentions (82% and 80%, respectively) in comparison to Pt/C (84%). Although the current density produced by these carbon electrocatalysts is still not on par with the commercial catalyst, the short-term stability indicates that some positive aspects have already been developed and that further tailoring of the surface chemistry may lead to ORR performances similar to Pt/C. Samples $CG/CNT_{O,5Fe,N}$, and Pt/C were then tested for methanol poisoning (Figure S8) by the addition of 1.5 mL of methanol to the electrolyte at ca. 1000 s of chronoamperometric measurements. These analyses demonstrated that the synthesized carbon material's electroactivity does not deteriorate with methanol poisoning, in contrast with Pt/C, confirming the potential of these materials toward the ORR in a methanol fuel cell. Due to the importance of CNTs in the performance of the hybrids, slightly increasing their amount during the hydrothermal process may lead to even higher electroactivities toward the ORR. Moreover, the incorporation of single atoms instead of using two different precursors may also improve the performance of the electrocatalysts and reduce their overall synthesis time and production costs. Therefore, the hybrids prepared in this work, although still underperforming compared to the Pt/C benchmark electrocatalyst, provide a good starting point for the development of green, sustainable, and cheap ORR electrocatalysts.

4. Conclusions

Carbon hybrids prepared by the hydrothermal treatment of glucose in the presence of oxidized CNTs were tailored by incorporating iron and/or nitrogen into their surface using two different methodologies: (i) the incorporation of iron followed by nitrogen doping and (ii) the simultaneous incorporation of the transition metal and the heteroatom in a single thermal treatment. The presence of CNT_O favored the polymerization of glucose during the hydrothermal treatment, resulting in materials with a lower concentration of functional groups than their glucose-derived carbon counterparts. Nitrogen doping of the monometallic samples results in materials with a much higher concentration of nitrogen than the Fe,N samples prepared using a single thermal treatment. The use of aH_2 atmosphere during the thermal decomposition and longer residence time for the decomposition of melamine leads to lower contents of N-6 and higher amounts of N-Q. Regarding the electrochemical performance, E_{onset} increases by 130 mV in comparison to the carbonized glucose sample due to the presence of active centers such as iron and nitrogen functional groups, whereas the current density increases linearly with the iron content. The reaction mechanism shifts to a direct four-electron mechanism by the addition of nitrogen functionalities to the hybrid, which decreases the formation of HO_2^- to 25%. The short-term durability of the carbon hybrids is higher than that of the glucose-derived carbon and increases due to the incorporation of iron and N-Q moieties. Therefore, considering the overall ORR performance parameters, the Fe,N carbon hybrid prepared using the one-step method presents the highest short-term stability (80%) and lowest formation of HO_2^- (27%), whereas the hybrid prepared by the two-step method exhibits the highest onset potential (0.83 V) and current density (2.95 mA cm^{-2}). Overall, Fe,N carbon hybrids present different advantages that may prove useful as a starting point to develop new electrocatalysts to replace the Pt/C benchmark.

Supplementary Materials: The following supporting information can be downloaded at: https://www.mdpi.com/article/10.3390/c10020047/s1, Figure S1: XPS survey spectra of the prepared Fe,N CG (a,b) and Fe,N CG/CNT hybrids (c,d) electrocatalysts, in which N-CG_{5Fe} and N-$CG/CNT_{O,5Fe}$ were prepared using a two-step method and $CG_{5Fe,N}$ and CG/CNT_O were prepared using a one-step method; Table S1: Surface chemical composition determined by XPS; Figure S2: N_2 adsorption–desorption isotherms at −196 °C and pore size distribution of undoped, CG_{XFe} and $CG/CNT_{O,5Fe}$ (a,c), and N-CG_{5Fe}, N-$CG/CNT_{O,5Fe}$, $CG_{5Fe,N}$ and $CG/CNT_{O,5Fe,N}$ (b,d) electrocatalysts; Figure S3: Elemental mapping of N-$CG/CNT_{O,5Fe}$ (a) and $CG/CNT_{O,5Fe,N}$ (b). Red—Carbon, Yellow—Iron, Green—Nitrogen, Blue—Oxygen; Figure S4: Raman spectra of undoped and monometallic electrocatalysts (a), and Fe,N electrocatalysts (b); Table S2: Ratio of the integrated D and G peaks obtained by Raman spectroscopy; Figure S5: Cyclic voltammograms recorded at 5 mV s^{-1} in N_2- and O_2-saturated 0.1 mol L^{-1} KOH electrolyte from 1.15 V to −0.05 V; Figure S6: Nyquist plots for CG and CG/CNT_O obtained by electrochemical impedance spectroscopy at 0 V; Figure S7: Onset potential (a) and current density at a potential of 0.4 V (b) obtained from the linear sweep voltammetry curves of Figure 5a (a) for the CG_{XFe} samples versus the percentage of Fe calculated by ICP; Table S3: ORR parameters of glucose-derived carbons, reagents and synthesis methods employed; Figure S8: Methanol tolerance of $CG/CNT_{O,5Fe,N}$ and Pt/C at 0.4 V during 10,000 s.

Author Contributions: Conceptualization, N.R.-R., J.L.F. and M.F.R.P.; methodology, R.G.M. and N.R.-R.; validation, R.G.M.; writing—original draft preparation, R.G.M. and N.R.-R.; writing—review and editing, J.L.F. and M.F.R.P.; visualization, N.R.-R.; supervision, N.R.-R. and M.F.R.P.; funding acquisition, M.F.R.P. All authors have read and agreed to the published version of the manuscript.

Funding: This work is financially supported by national funds through the FCT/MCTES (PIDDAC), under the project Bicat4Energy, PTDC/EQU-EQU/1707/2020. This research was also funded by FCT/MCTES (PIDDAC): LSRE-LCM, UIDB/50020/2020 (https://doi.org/10.54499/UIDB/50020/2020) and UIDP/50020/2020 (https://doi.org/10.54499/UIDP/50020/2020); ALiCE, LA/P/0045/2020 (https://doi.org/10.54499/LA/P/0045/2020). RGM acknowledges the Research Grant from FCT 2020.06422.BD (https://doi.org/10.54499/2020.06422.BD).

Institutional Review Board Statement: Not applicable.

Informed Consent Statement: Not applicable.

Data Availability Statement: The raw data supporting the conclusions of this article will be made available by the authors on request.

Acknowledgments: The authors are indebted to CEMUP for assistance with SEM, BSED, EDS and elemental mapping analyses.

Conflicts of Interest: The authors declare no conflicts of interest.

References

1. Kirubakaran, A.; Jain, S.; Nema, R. A review on fuel cell technologies and power electronic interface. *Renew. Sustain. Energy Rev.* **2009**, *13*, 2430–2440. [CrossRef]
2. Stacy, J.; Regmi, Y.N.; Leonard, B.; Fan, M. The recent progress and future of oxygen reduction reaction catalysis: A review. *Renew. Sustain. Energy Rev.* **2017**, *69*, 401–414. [CrossRef]
3. Goswami, C.; Hazarika, K.K.; Bharali, P. Transition metal oxide nanocatalysts for oxygen reduction reaction. *Mater. Sci. Energy Technol.* **2018**, *1*, 117–128. [CrossRef]
4. Yang, J.; Toshimitsu, F.; Yang, Z.; Fujigaya, T.; Nakashima, N. Pristine carbon nanotube/iron phthalocyanine hybrids with a well-defined nanostructure show excellent efficiency and durability for the oxygen reduction reaction. *J. Mater. Chem. A* **2017**, *5*, 1184–1191. [CrossRef]
5. Komba, N.; Zhang, G.; Wei, Q.; Yang, X.; Prakash, J.; Chenitz, R.; Rosei, F.; Sun, S. Iron (II) phthalocyanine/N-doped graphene: A highly efficient non-precious metal catalyst for oxygen reduction. *Int. J. Hydrogen Energy* **2019**, *44*, 18103–18114. [CrossRef]
6. Hyun, K.; Ueno, T.; Panomsuwan, G.; Li, O.L.; Saito, N. Heterocarbon nanosheets incorporating iron phthalocyanine for oxygen reduction reaction in both alkaline and acidic media. *Phys. Chem. Chem. Phys.* **2016**, *18*, 10856–10863. [CrossRef]
7. Han, S.; Hu, X.; Wang, J.; Fang, X.; Zhu, Y. Novel Route to Fe-Based Cathode as an Efficient Bifunctional Catalysts for Rechargeable Zn-Air Battery. *Adv. Energy Mater.* **2018**, *8*, 1800955. [CrossRef]
8. Chung, H.T.; Cullen, D.A.; Higgins, D.; Sneed, B.T.; Holby, E.F.; More, K.L.; Zelenay, P. Direct atomic-level insight into the active sites of a high-performance PGM-free ORR catalyst. *Science* **2017**, *357*, 479–484. [CrossRef]
9. Zhao, S.; Wang, D.; Amal, R.; Dai, L. Carbon-Based Metal-Free Catalysts for Key Reactions Involved in Energy Conversion and Storage. *Adv. Mater.* **2019**, *31*, e1801526.
10. Figueiredo, J.L. Nanostructured porous carbons for electrochemical energy conversion and storage. *Surf. Coat. Technol.* **2018**, *350*, 307–312. [CrossRef]
11. Wu, Z.; Song, M.; Wang, J.; Liu, X. Recent Progress in Nitrogen-Doped Metal-Free Electrocatalysts for Oxygen Reduction Reaction. *Catalysts* **2018**, *8*, 196. [CrossRef]
12. Wang, J.; Wang, H.; Fan, Y. Techno-Economic Challenges of Fuel Cell Commercialization. *Engineering* **2018**, *4*, 352–360. [CrossRef]
13. He, C.; Sankarasubramanian, S.; Matanovic, I.; Atanassov, P.; Ramani, V. Understanding the Oxygen Reduction Reaction Activity and Oxidative Stability of Pt Supported on Nb-Doped TiO_2. *ChemSusChem* **2019**, *12*, 3468–3480. [CrossRef]
14. Li, Z.; Zhou, P.; Zhao, Y.; Jiang, W.; Zhao, B.; Chen, X.; Li, M. Ultradurable Pt-Based Catalysts for Oxygen Reduction Electrocatalysis. *Catalysts* **2024**, *14*, 57. [CrossRef]
15. Lee, K.; Zhang, L.; Lui, H.; Hui, R.; Shi, Z.; Zhang, J. Oxygen reduction reaction (ORR) catalyzed by carbon-supported cobalt polypyrrole (Co-PPy/C) electrocatalysts. *Electrochim. Acta* **2009**, *54*, 4704–4711. [CrossRef]
16. Paul, R.; Dai, Q.; Hu, C.; Dai, L. Ten years of carbon-based metal-free electrocatalysts. *Carbon Energy* **2019**, *2*, 151–152. [CrossRef]

17. Yan, L.; Yu, J.; Houston, J.; Flores, N.; Luo, H. Biomass derived porous nitrogen doped carbon for electrochemical devices. *Green Energy Environ.* **2017**, *2*, 84–99. [CrossRef]
18. Jain, A.; Balasubramanian, R.; Srinivasan, M.P. Hydrothermal conversion of biomass waste to activated carbon with high porosity: A review. *Chem. Eng. J.* **2016**, *283*, 789–805. [CrossRef]
19. Morais, R.G.; Rey-Raap, N.; Costa, R.S.; Pereira, C.; Guedes, A.; Figueiredo, J.L.; Pereira, M.F.R. Hydrothermal Carbon/Carbon Nanotube Composites as Electrocatalysts for the Oxygen Reduction Reaction. *J. Compos. Sci.* **2020**, *4*, 20. [CrossRef]
20. Morais, R.G.; Rey-Raap, N.; Figueiredo, J.L.; Pereira, M.F.R. Glucose-derived carbon materials with tailored properties as electrocatalysts for the oxygen reduction reaction. *Beilstein J. Nanotechnol.* **2019**, *10*, 1089–1102. [CrossRef]
21. John, K.I.; Omorogie, M.O. Biomass-based hydrothermal carbons for catalysis and environmental cleanup: A review. *Green Chem. Lett. Rev.* **2022**, *15*, 162–186. [CrossRef]
22. Skorupska, M.; Ilnicka, A.; Lukaszewicz, J.P. Successful Manufacturing Protocols of N-Rich Carbon Electrodes Ensuring High ORR Activity: A Review. *Processes* **2022**, *10*, 643. [CrossRef]
23. Sathiskumar, C.; Ramakrishnan, S.; Vinothkannan, M.; Kim, A.R.; Karthikeyan, S.; Yoo, D.J. Nitrogen-Doped Porous Carbon Derived from Biomass Used as Trifunctional Electrocatalyst toward Oxygen Reduction, Oxygen Evolution and Hydrogen Evolution Reactions. *Nanomaterials* **2020**, *10*, 76. [CrossRef]
24. Choi, E.Y.; Kim, D.E.; Lee, S.Y.; Kim, C.K. Electrocatalytic activity of nitrogen-doped holey carbon nanotubes in oxygen reduction and evolution reactions and their application in rechargeable zinc–air batteries. *Carbon* **2020**, *166*, 245–255. [CrossRef]
25. Liu, A.; Ma, M.; Zhang, X.; Ming, J.; Jiang, L.; Li, Y.; Zhang, Y.; Liu, S. A biomass derived nitrogen doped carbon fibers as efficient catalysts for the oxygen reduction reaction. *J. Electroanal. Chem.* **2018**, *824*, 60–66. [CrossRef]
26. Barrera, D.; Florent, M.; Kulko, M.; Bandosz, T.J. Ultramicropore-influenced mechanism of oxygen electroreduction on metal-free carbon catalysts. *J. Mater. Chem. A* **2019**, *7*, 27110–27123. [CrossRef]
27. Rey-Raap, N.; Enterría, M.; Martins, J.I.; Pereira, M.F.R.; Figueiredo, J.L. Influence of Multiwalled Carbon Nanotubes as Additives in Biomass-Derived Carbons for Supercapacitor Applications. *ACS Appl. Mater. Interfaces* **2019**, *11*, 6066–6077. [CrossRef]
28. Wang, W.; Jia, Q.; Mukerjee, S.; Chen, S. Recent Insights into the Oxygen-Reduction Electrocatalysis of Fe/N/C Materials. *ACS Catal.* **2019**, *9*, 10126–10141. [CrossRef]
29. Ma, Q.; Jin, H.; Zhu, J.; Li, Z.; Xu, H.; Liu, B.; Zhang, Z.; Ma, J.; Mu, S. Stabilizing Fe–N–C Catalysts as Model for Oxygen Reduction Reaction. *Adv. Sci.* **2021**, *8*, 2102209. [CrossRef]
30. Mamtani, K.; Jain, D.; Co, A.C.; Ozkan, U.S. Nitrogen-Coordinated Iron—Carbon as Efficient Bifunctional Electrocatalysts for the Oxygen Reduction and Oxygen Evolution Reactions in Acidic Media. *Energy Fuels* **2017**, *31*, 6541–6547. [CrossRef]
31. Yan, Z.; Dai, C.; Zhang, M.; Lv, X.; Zhao, X.; Xie, J. Nitrogen doped porous carbon with iron promotion for oxygen reduction reaction in alkaline and acidic media. *Int. J. Hydrogen Energy* **2019**, *44*, 4090–4101. [CrossRef]
32. Liu, Y.; Huang, B.; Zhang, X.; Huang, X.; Xie, Z. In-situ fabrication of nitrogen-doped carbon nanosheets containing highly dispersed single iron atoms for oxygen reduction reaction. *J. Power Sources* **2019**, *412*, 125–133. [CrossRef]
33. Abdelwahab, A.; Farghali, A.A.; Allah, A.E. Synergy between iron oxide sites and nitrogen-doped carbon xerogel/diamond matrix for boosting the oxygen reduction reaction. *Nanoscale Adv.* **2022**, *4*, 837–848. [CrossRef]
34. Xia, J.; Li, K.; Zhao, G.-L. Electric Field Polarization to Increase Bifunctional Oxygen Electrocatalyst Performance of Nitrogen–Iron Functionalized Carbon Nanomaterials. *ACS Appl. Energy Mater.* **2020**, *3*, 1484–1495. [CrossRef]
35. Canal-Rodríguez, M.; Rey-Raap, N.; Menéndez, J.Á.; Montes-Morán, M.A.; Figueiredo, J.L.; Pereira, M.F.R.; Arenillas, A. Effect of porous structure on doping and the catalytic performance of carbon xerogels towards the oxygen reduction reaction. *Micropor. Mesopor. Mater.* **2020**, *293*, 109811. [CrossRef]
36. Morais, R.; Rey-Raap, N.; Figueiredo, J.; Pereira, M. Highly electroactive N–Fe hydrothermal carbons and carbon nanotubes for the oxygen reduction reaction. *J. Energy Chem.* **2020**, *50*, 260–270. [CrossRef]
37. Ba, H.; Duong-Viet, C.; Liu, Y.; Nhut, J.-M.; Granger, P.; Ledoux, M.J.; Pham-Huu, C. Pham-Huu, Nitrogen-doped carbon nanotube spheres as metal-free catalysts for the partial oxidation of H_2S. *Comptes Rendus Chim.* **2016**, *19*, 1303–1309. [CrossRef]
38. Enterría, M.; Figueiredo, J.L. Nanostructured mesoporous carbons: Tuning texture and surface chemistry. *Carbon* **2016**, *108*, 79–102. [CrossRef]
39. Lai, L.; Potts, J.R.; Zhan, D.; Wang, L.; Poh, C.K.; Tang, C.; Gong, H.; Shen, Z.; Lin, J.; Ruoff, R.S. Exploration of the active center structure of nitrogen-doped graphene-based catalysts for oxygen reduction reaction. *Energy Environ. Sci.* **2012**, *5*, 7936–7942. [CrossRef]
40. Kecili, R.; Hussain, C.M. Chapter 4—Mechanism of Adsorption on Nanomaterials. In *Nanomaterials in Chromatography*; Hussain, C.M., Ed.; Elsevier: Amsterdam, The Netherlands, 2018; pp. 89–115.
41. Chang, S.-S.; Clair, B.; Ruelle, J.; Beauchêne, J.; Di Renzo, F.; Quignard, F.; Zhao, G.-J.; Yamamoto, H.; Gril, J. Mesoporosity as a new parameter for understanding tension stress generation in trees. *J. Exp. Bot.* **2009**, *60*, 3023–3030. [CrossRef]
42. Yurdakal, S.; Garlisi, C.; Özcan, L.; Bellardita, M.; Palmisano, G. Chapter 4—(Photo)catalyst Characterization Techniques: Adsorption Isotherms, SEM, FTIR, UV-VIS, Photoluminescence and Electrochemical Characterizations. In *Heterogeneous Photocatalysis*; Marcì, G., Palmisano, L., Eds.; Elsevier: Amsterdam, The Netherlands, 2019; pp. 87–152.
43. Bedia, J.; Monsalvo, V.; Rodriguez, J.; Mohedano, A. Iron catalysts by chemical activation of sewage sludge with $FeCl_3$ for CWPO. *Chem. Eng. J.* **2017**, *318*, 224–230. [CrossRef]

44. Falco, C.; Marco-Lozar, J.; Salinas-Torres, D.; Morallón, E.; Cazorla-Amorós, D.; Titirici, M.; Lozano-Castelló, D. Tailoring the porosity of chemically activated hydrothermal carbons: Influence of the precursor and hydrothermal carbonization temperature. *Carbon* **2013**, *62*, 346–355. [CrossRef]
45. Sevilla, M.; Fuertes, A.B. Chemical and Structural Properties of Carbonaceous Products Obtained by Hydrothermal Carbonization of Saccharides. *Chem.-A Eur. J.* **2009**, *15*, 4195–4203. [CrossRef] [PubMed]
46. Xiao, L.; Schroeder, M.; Kluge, S.; Balducci, A.; Hagemann, U.; Schulz, C.; Wiggers, H. Direct self-assembly of Fe_2O_3/reduced graphene oxide nanocomposite for high-performance lithium-ion batteries. *J. Mater. Chem. A* **2015**, *3*, 11566–11574. [CrossRef]
47. Chen, W.; Xiang, Q.; Peng, T.; Song, C.; Shang, W.; Deng, T.; Wu, J. Reconsidering the Benchmarking Evaluation of Catalytic Activity in Oxygen Reduction Reaction. *iScience* **2020**, *23*, 101532. [CrossRef] [PubMed]
48. Shinagawa, T.; Garcia-Esparza, A.T.; Takanabe, K. Insight on Tafel slopes from a microkinetic analysis of aqueous electrocatalysis for energy conversion. *Sci. Rep.* **2015**, *5*, 13801. [CrossRef]
49. Ma, R.; Lin, G.; Zhou, Y.; Liu, Q.; Zhang, T.; Shan, G.; Yang, M.; Wang, J. A review of oxygen reduction mechanisms for metal-free carbon-based electrocatalysts. *NPJ Comput. Mater.* **2019**, *5*, 78. [CrossRef]
50. Peng, Y.; Liu, H. Effects of Oxidation by Hydrogen Peroxide on the Structures of Multiwalled Carbon Nanotubes. *Ind. Eng. Chem. Res.* **2006**, *45*, 6483–6488. [CrossRef]
51. Singh, H.; Zhuang, S.; Ingis, B.; Nunna, B.B.; Lee, E.S. Carbon-based catalysts for oxygen reduction reaction: A review on degradation mechanisms. *Carbon* **2019**, *151*, 160–174. [CrossRef]
52. Morais, R.; Rey-Raap, N.; Figueiredo, J.; Pereira, M. Fe, Co, N-doped carbon nanotubes as bifunctional oxygen electrocatalysts. *Appl. Surf. Sci.* **2022**, *572*, 151459. [CrossRef]

C Journal of Carbon Research

MDPI

Article

Algae Derived Carbon from Hydrothermal Liquefaction as Sustainable Carbon Electrode Material for Supercapacitor

Kingsford Asare [1], Abhijeet Mali [1], Md Faruque Hasan [1], Philip Agbo [1], Abolghasem Shahbazi [2,*] and Lifeng Zhang [1,*]

1 Department of Nanoengineering, Joint School of Nanoscience and Nanoengineering, North Carolina A&T State University, Greensboro, NC 27401, USA
2 Department of Natural Resources and Environmental Design, College of Agriculture and Environmental Sciences, North Carolina A&T State University, 1601 E Market St., Greensboro, NC 27411, USA
* Correspondence: ash@ncat.edu (A.S.); lzhang@ncat.edu (L.Z.)

Abstract: With the worldwide awareness of sustainability, biomass-derived carbon electrode materials for supercapacitors have attracted growing attention. In this research, for the first time, we explored the feasibility of making use of the carbon byproduct from hydrothermal liquefaction (HTL) of microalgae, termed herein as algae-derived carbon (ADC), to prepare sustainable carbon electrode materials for high-performance supercapacitor development. Specifically, we investigated carbon activation with a variety of activating reagents as well as N- and Fe-doping of the obtained ADC with the intention to enhance its electrochemical performance. We characterized the structure of the activated and doped ADCs using scanning electron microscope (SEM), Raman spectroscopy, Fourier transform infrared spectroscopy (FTIR), X-ray photoelectron spectroscopy (XPS), and BET surface area and pore analysis, and correlated the ADCs' structure with their electrochemical performance as evaluated using cyclic voltammetry (CV), galvanostatic charge/discharge (GCD), impedance, and cycle stability through an assembled symmetric two-electrode cell with 1 M H_2SO_4 as electrolyte. It was found that the ADC that is activated using KOH (KOH-ADC) showed the best electrochemical performance, and its specific capacitance was 14.1-fold larger with respect to that of the raw ADC and reached 234.5 F/g in the GCD test at a current density of 0.5 A/g. The KOH-ADC also demonstrated excellent capacitance retention (97% after 10,000 cycles at a high current density of 10 A/g) for stable long-term operations. This research pointed out a promising direction to develop sustainable electrode materials for supercapacitors from the carbon byproduct produced after HTL processing of algae.

Keywords: algae; sustainable carbon; hydrothermal liquefaction; electrode material; supercapacitor

1. Introduction

With the awareness of sustainable development, increasing research efforts have been channeled towards developing clean, efficient, and sustainable sources of energy and energy storage devices to meet the energy demand of the human society in the 21st century [1]. Among the various energy storage devices that are being explored, supercapacitors have attracted great attention owing to their high power density, cycle stability, and fast charge and discharge rates [2]. Supercapacitors can be broadly classified into two main categories: electrochemical double layer capacitors (EDLC) and pseudocapacitors. The mechanism of charge storage in EDLC is based on a non-faradaic reversible electrical charge transfer and storage at the electrode–electrolyte interface, while that of the pseudocapacitor is based on a faradaic redox reaction process [3]. Although the latter shows good electrochemical performance, it often suffers from issues of high manufacturing costs and toxicity [4]. On the contrary, EDLC supercapacitors have relatively lower manufacturing cost and environmental benignity. Additionally, EDLC supercapacitors exhibit a high cycle stability and can maintain their electrochemical performance even up to 10^6 cycles or more [5].

A wide range of electrode materials have been suggested for the development of supercapacitors. Carbon has been widely used for supercapacitor electrode materials because of its low cost, excellent electrical conductivity, and high chemical stability. Due to the drive towards sustainability, biomass-derived carbon electrode materials have become a direction of current research [6]. Hydrothermal liquefaction (HTL) is an important thermochemical process that generates bio-crude oil from biomass [7]. Along with the bio-crude oil product, it is noteworthy that the HTL processing of biomass also generates a significant amount of solid carbon residue (2–70 wt.% of the total product depending on feedstock) [8]. Up to date, these carbon residue byproducts are mostly being utilized for the purpose of carbon sequestering and wastewater treatment [9–11]. In a recent report, Shell et al. (2021) utilized the HTL process to develop carbon electrode materials from corn stover and obtained a specific capacitance of 242 F/g at 5 mV/s in the cyclic voltammetry test and an energy density of 9.9 W h/kg in 2 M KOH [12]. The findings from this study indicated the possibility of acquiring carbon electrode materials from the HTL processing of biomass for supercapacitor applications. However, current studies in developing sustainable carbon electrode materials for supercapacitor use lie in plant and animal biomass [13] and little attention has been given to algae, which represent a distinctive class of biomass. There is no report yet to employ carbon residue particles from the HTL processing of algae (termed as algae-derived carbon (ADC) herein) as a value-added product for supercapacitor electrodes, even though HTL is an important thermochemical process to generate bio-crude oil from algal biomass [14]. It is well known that carbon sources and processing techniques impact the quality of carbon electrode materials for supercapacitors [12,15–17]. The ADC could possess a rich supply of heteroatoms like nitrogen which comes from the protein component of algae and thus could contribute to pseudocapacitance as an electrode material for supercapacitors.

In this research, for the first time, we seek to specifically explore the potential of ADC as an electrode material for supercapacitor application. With the intention to enhance the electrochemical performance of the obtained ADC, we further performed carbon activation and doping studies, which are popular ways to improve the electrochemical performance of carbon electrode materials and has been proved in our previous research [18,19]. For the carbon activation study, we investigated H_2SO_4, H_3PO_4, HNO_3, and KOH according to their different activation mechanisms. Specifically, H_2SO_4 and HNO_3 can improve the amount of functional groups on the surface of ADC through the oxidation and nitration of aromatic rings in the ADC [20]; H_3PO_4 can perform a depolymerization, dehydration, and redistribution of constituent biopolymers to develop porous structures in biochar [21]; KOH can etch the carbon surface to create a porous carbon structure using a decarburization mechanism. It is well known that introducing heteroatoms, especially N, can improve the electrochemical performance of carbon electrode materials [22] and melamine is a good N source for doping purposes [23], while iron-doped carbon electrode materials also demonstrated enhanced electrochemical performance [24]. For the carbon doping study, we thus investigated the N-doping effect using melamine and the Fe-doping effect using iron (III) chloride. Overall, we correlated the electrochemical performance of our ADC electrode materials with the ADC structure from various activations and doping processes and revealed a feasible way to make use of the ADC from HTL processing as a value-added product for energy storage applications.

2. Materials and Methods

2.1. Materials

Algal biomass (*Chrollela vulgaris*) was acquired from nuts.com (Cranford, NJ, USA). Potassium hydroxide (85%), sulfuric acid (98%), carbon black, iron (III) chloride hexahydrate (99+%), melamine, N-methyl-2-pyrrolidone (NMP), and polyvinylidene fluoride (PVDF) were purchased from Sigma-Aldrich (St. Louis, MO, USA). Phosphoric acid (pure, 85+%) was purchased from Cole-Parmer (Vernon Hills, IL, USA) and nitric acid was bought from Thermo Fisher Scientific (Waltham, MA, USA). All the chemicals were used as purchased without any further purification.

2.2. Preparation of Algae-Derived Carbon (ADC)

In a typical HTL process to prepare ADC, 73 mg of raw algae was mixed with 700 mL DI water and placed in a pressure Parr reactor (model: 4843). The algae system was subjected to a reaction temperature of 300 °C at a pressure of 1500 psi for 30 min. After the reaction, the solid residue, i.e., ADC particles, was separated from the HTL product, washed, and dried in an oven at 60 °C.

To prepare activated and doped ADCs, in the case of H_2SO_4, H_3PO_4, and HNO_3, the raw ADC particles were immersed in the corresponding acid overnight and then the particles were filtered, washed, and dried for heat treatment; in the case of melamine (N source), iron (III) chloride hexahydrate (Fe source), and KOH, the raw ADC particles were mixed with these chemicals at a mixing ratio of 1:1 (w/w), respectively. All the ADCs were subsequently heated in a furnace at 800 °C for 1 h in a nitrogen environment. The obtained respective activated and doped ADCs were washed with deionized water until neutrality and then dried in oven at 60 °C. The as-prepared ADCs were then used for electrode preparation, respectively. In the following part, individual ADC samples are correspondingly labeled as Raw-ADC, H_2SO_4-ADC, H_3PO_4-ADC, HNO_3-ADC, N-ADC, Fe-ADC, and KOH-ADC, respectively.

2.3. Material Characterization

The surface morphology of all the ADC materials was revealed by using a field emission SEM (Zeiss Auriga FIB FESEM, White Plains, NY, USA). Raman spectroscopy was performed to characterize the carbon structure of the ADCs using a Horiba Raman confocal microscope (Piscataway, NJ, USA) with an excitation wavelength of 532 nm. FTIR spectroscopy was carried out by employing an Agilent 670 FTIR spectrometer w/ATR (Santa Clara, CA, USA) to identify functional groups and covalent bonds that are present in ADCs. Specific surface area and porosity characteristics of the ADCs were investigated using a Micrometrics ASAP 2020 surface area and porosity system analyzer (Norcross, GA, USA). Surface elemental information of the ADCs was characterized using a Thermo Scientific Escalab Xi+ X-ray photon spectrometer (XPS, Waltham, MA, USA). In the XPS measurements, surface charge effects were minimized and all XPS spectra were charge-corrected using C_{1s} binding energy of 284.8 eV.

2.4. Electrode Preparation and Characterization

The supercapacitor electrodes were prepared by mixing respective as-prepared ADC particles, carbon black, and polyvinylidene fluoride (PVDF) at a mixing ratio of 80:10:10. A few drops of N-methyl-2-pyrrolidone (NMP) solution were added to the ADC mixture to form a slurry. The slurry was cast onto a stainless steel mesh and dried in an oven at 60 °C overnight. This dried paste was then pressed using a 1.5 metric ton load to obtain a flat electrode for the supercapacitor test. A symmetric two-electrode cell, which consisted of the respective electrode material, the separator (filter paper), and 1 M H_2SO_4 electrolyte, was used to carry out electrochemical analyses including cyclic voltammetry (CV), galvanostatic charge/discharge (GCD), impedance, and cycle stability tests using a CHI660E electrochemical workstation to evaluate the electrochemical performance of the respective electrode materials. The impedances of the cell were carried out in the frequency range of 100 kHz and 0.1 Hz.

The specific capacitance was computed using Equations (1) and (2).

$$\text{From the CV test}: \ C = \frac{1}{mv(V_b - V_a)} \int_{V_a}^{V_b} IdV \tag{1}$$

$$\text{From the GCD test}: \ C = \frac{2 \times I \times \Delta t}{m \times \Delta V} \tag{2}$$

In Equation (1), C (F/g) is the specific capacitance; m (g) is the mass of active material loaded in the working electrode; v (V/s) is the scan rate; I (A) is the current; V_b and V_a (V) are the upper and lower potential limits of the chosen potential window. In Equation (2), I (A) is the discharge current; Δt (s) is the discharge time; ΔV (V) is the potential window; and m (g) is the active mass of a single electrode.

The energy density of the cell (E, W h/kg) was computed using Equation (3):

$$E = \frac{1}{8 \times 3.6} \times C \times \Delta V^2 \tag{3}$$

where C is the specific capacitance of the electrode material from the GCD test and ΔV is the potential window.

The power density of the cell (P, W/kg) was computed using Equation (4):

$$P = \frac{3600 \times E}{t} \tag{4}$$

where E is the energy density of the cell and t is the discharge time.

3. Results and Discussion

3.1. Morphology

The obtained Raw-ADC comprised of particles with sizes from a few micrometers to hundreds of micrometers (Figure 1A). The surface of the Raw-ADC particles was relatively smooth (Figure 1B). All the activated and doped ADCs showed a relatively rough surface with broken and finer particles (Figure 1C–H). It is noteworthy that KOH-ADC presented the most surface roughness with well-developed porosity (Figure 1H). This is a result of the high carbon etching effect that is associated with KOH.

Figure 1. *Cont.*

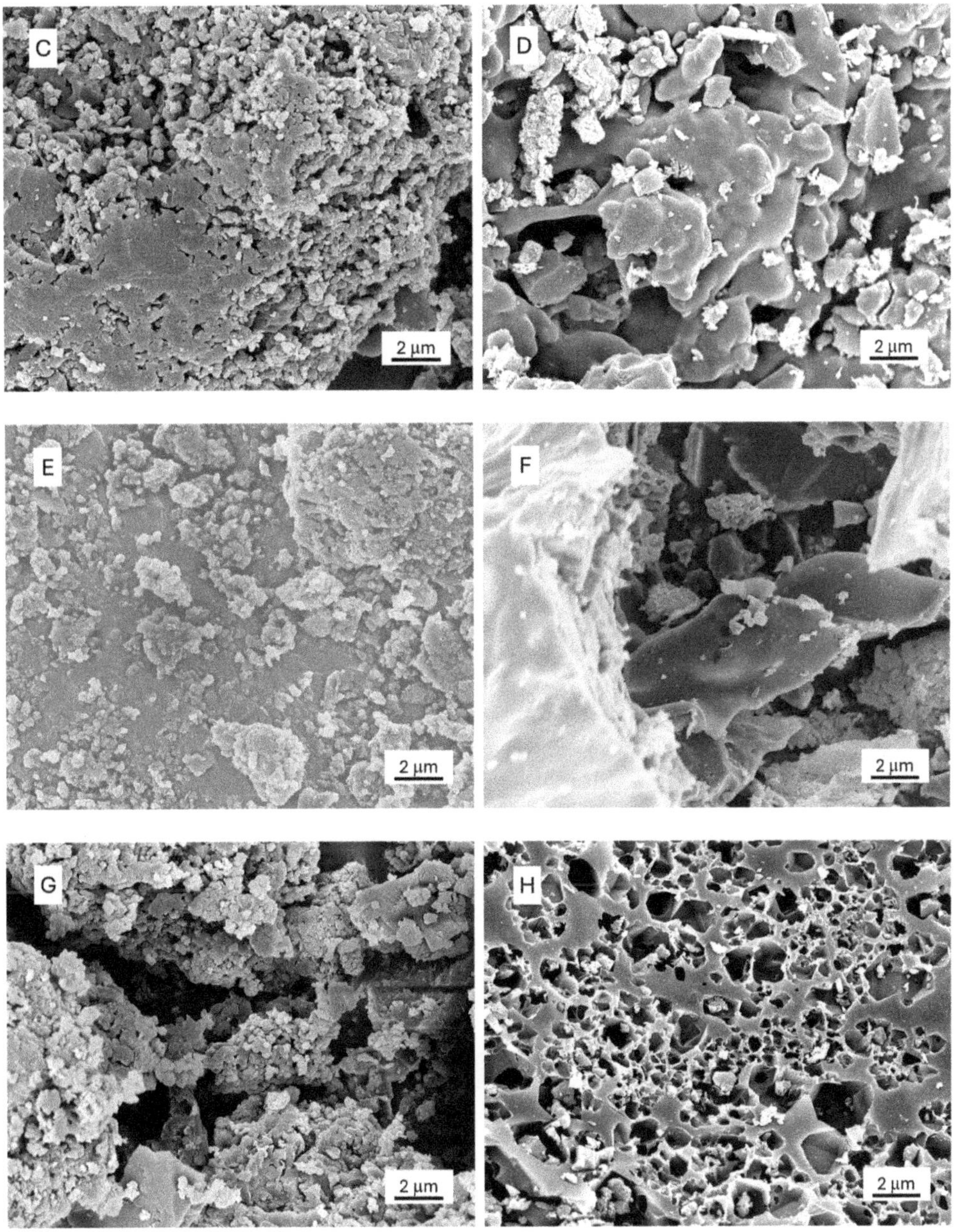

Figure 1. SEM images of Raw-ADC (low magnification (**A**) and high magnification (**B**)), H_2SO_4-ADC (**C**), H_3PO_4-ADC (**D**), HNO_3-ADC (**E**), N-ADC (**F**), Fe-ADC (**G**), and KOH-ADC (**H**).

3.2. Structure

Raman spectroscopy was used to study the carbon structure of all the ADC samples (Figure 2). In Raman spectroscopy, the band between 1598 and 1575 cm^{-1} (G-band) is attributed to the organized carbon structure, while the band between 1355 and 1330 cm^{-1} (D-band) is attributed to the disordered carbon structure. Thus, a higher intensity ratio of

I_G/I_D indicates more sp^2 hybridized ordered carbon structures in corresponding carbon materials. Among all the studied ADCs, the Raw-ADC exhibited the highest I_G/I_D ratio of 1.47, implying the presence of the highest level of the ordered graphitic phase in the Raw-ADC. Compared to the Raw-ADC, the activation and doping resulted in reduced I_G/I_D ratios, i.e., more defective carbon structures, with N-ADC having the lowest I_G/I_D ratio of 0.85, showing the most defective carbon structure. Basically, the defective carbon structure is a result of carbon etching caused by activation and doping.

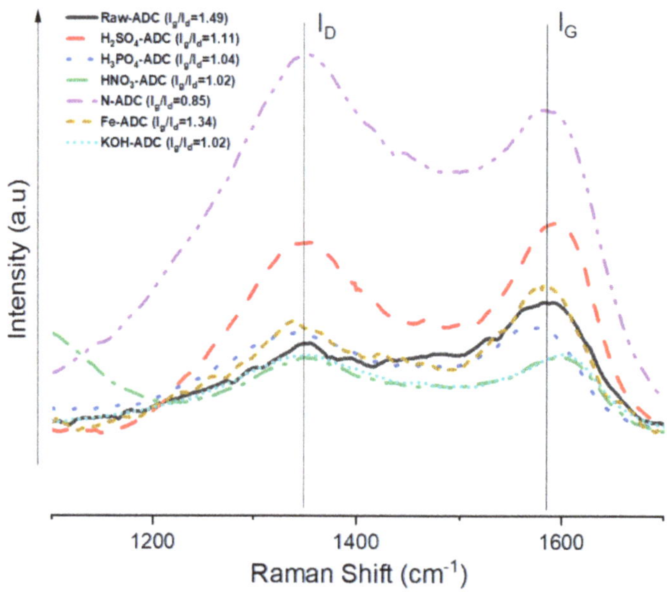

Figure 2. Raman spectroscopy of all the ADC materials.

FTIR analysis was carried out to study the chemical functional groups of the as-synthesized ADC (Raw-ADC in Figure 3). The Raw-ADC showed a broad and strong IR peak around 3375 cm^{-1}, which can be attributed to the stretching vibrations of O–H and/or N–H. The peaks observed between 2800 and 3000 cm^{-1}, as well as 1460 and 1380 cm^{-1}, represent C–H vibrations. The relative broad peaks observed around 1630 cm^{-1} and 1560 cm^{-1} could be attributed to Amide I and II and/or aromatic C=C stretching vibrations. The peaks in the range of 1100 and 1170 cm^{-1} could be attributed to C–O–C and/or C–O stretching vibrations. The peaks in the range of 1300 and 1350 cm^{-1} could be attributed to aromatic C–N stretching and/or O–H bending. After activation and doping, except for Fe-ADC, all the ADC materials showed new peaks in the range of 1950–2200 cm^{-1} (Figure 3), indicating oxidized functional groups like C≡C, C=C=C, C=C=O, and C=C=N, as well as aromatic compounds. This could be attributed to the oxidation and removal of amorphous carbon structures.

XPS was further carried out to analyze the surface element composition and bonding of all the ADC materials (Figure 4 and Table 1). The full XPS survey indicated that Raw-ADC already contained a significant amount of the N element. Activation and doping generally increased the amount of the N element, except for HNO_3 and KOH. Interestingly, Fe doping resulted in the largest amount of the N element. Particularly, KOH activation led to the smallest amount of the N element, the largest amount of the O element, and the smallest amount of the C element. Since N bonding, including pyridinic N, pyrrolic N, quaternary N, and oxidized N, can have significant effects on the electrochemical performance of the carbon electrode material in supercapacitors [18], high-resolution N_{1s} XPS spectra were obtained and analyzed (Table 2 and Figure S1). The N_{1s} peak of ADCs can be deconvoluted into four peaks at 398.5 ± 0.5, 399.5 ± 0.5, 400.5 ± 0.5, and 403 ± 0.5 eV, which correspond to pyridinic N, pyrrolic N, quaternary N, and oxidized N, respectively [18]. It was found that the Raw-ADC did not contain any pyrrolic N, while KOH-ADC demonstrated the highest relative proportion of pyrrolic-N (85%), followed by HNO_3-ADC.

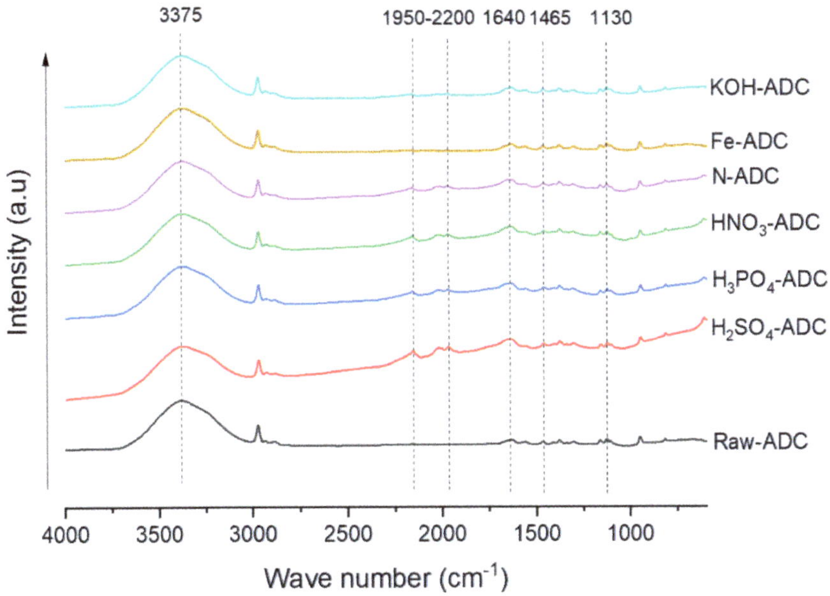

Figure 3. FTIR spectroscopy of all the ADC materials.

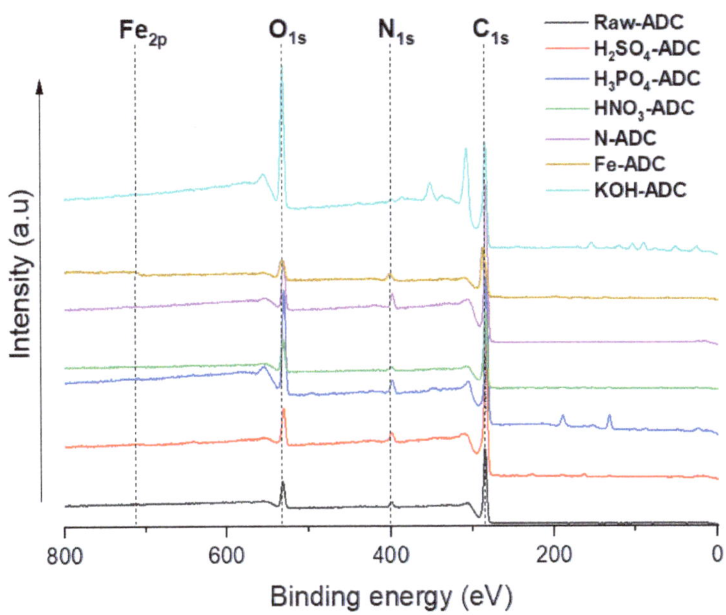

Figure 4. XPS spectra of all the ADC materials.

Table 1. Relative atomic percentage of selected elements present in the ADC materials.

ADC Materials	C (Atom%)	O (Atom%)	N (Atom%)	Fe (Atom%)	C/N
Raw-ADC	84.03	12.21	3.76	-	22.34
H_2SO_4-ADC	85.7	9.12	5.19	-	16.51
H_3PO_4-ADC	73.1	21.7	5.2	-	14.06
HNO_3-ADC	81.82	15.1	3.08	-	26.56
N-ADC	83.93	11.01	5.06	-	16.59
Fe-ADC	79.33	13.35	6.98	0.33	11.37
KOH-ADC	63.43	35.05	1.52	-	41.73

Table 2. Area percentage of the deconvoluted high-resolution XPS N_{1s} peaks of the ADC materials.

ADC Materials	Pyridinic-N (%)	Pyrrolic-N (%)	Quaternary-N (%)	Oxidized-N (%)
Raw-ADC	100			
H_2SO_4-ADC	41.35	6.72	51.93	
H_3PO_4-ADC	100			
HNO_3-ADC	27.91	51.83	8.84	11.43
N-ADC	100			
Fe-ADC	74.23		17.58	8.18
KOH-ADC		85.13		14.87

The specific surface area and porosity characteristics of all the ADCs were obtained from the BET surface area analysis (Figure 5 and Table 3). All the ADCs displayed type H4 hysteresis loops, implying the existence of micro- and meso-porous structures. The Raw-ADC, however, did not contain any micropores and showed a low specific surface area of 5.85 m^2/g, indicating limited porous structure. Compared to the Raw-ADC, the processes of activation and doping increased micropore and mesopore volumes, as well as the specific surface area, which is consistent with the break of particles and surface roughness as observed from SEM. Among all the ADC materials, H_3PO_4 activation and melamine treatment for N doping resulted in the least pore development, while KOH-ADC exhibited a marvelous pore structure development with the smallest average pore size (2.20 nm), the largest pore volume (0.1716 cm^3/g), and the largest specific surface area (1418 m^2/g), indicating the most efficient carbon etching and consequent pore formation. The largest micropore and mesopore volume from KOH activation could be attributed to the CO or CO_2 generation [25], as well as the intercalation mechanism by metallic K within the carbon framework, leading to the widening of pores during the high temperature decomposition process [26].

Table 3. BET surface area and porosity of all the ADC materials.

ADC Materials	BET SSA	V_{micro}	V_{meso}	V_{total}	Average Pore Size
	(m^2/g)	(cm^3/g)	(cm^3/g)	(cm^3/g)	(nm)
Raw-ADC	5.85	0.0000	0.0065	0.0487	33.51
H_2SO_4-ADC	166.21	0.0001	0.0234	0.0484	3.10
H_3PO_4-ADC	71.98	0.0048	0.0338	0.0593	4.27
HNO_3-ADC	152.69	0.0047	0.0315	0.0586	3.17
N-ADC	65.81	0.0027	0.0174	0.03003	3.37
Fe-ADC	174.66	0.0037	0.0173	0.0282	2.45
KOH-ADC	1417.92	0.0587	0.0975	0.1716	2.20

SSA—specific surface area; V_{micro}—micropore volume; V_{meso}—mesopore volume; V_{total}—total pore volume.

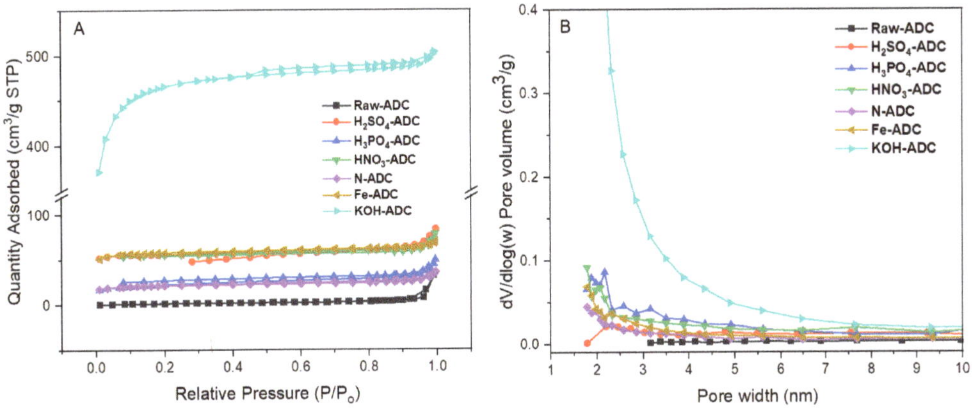

Figure 5. N_2 adsorption/desorption isotherms (**A**) and pore size distribution (**B**) of all the ADC materials.

3.3. Electrochemical Properties

Electrochemical studies including CV and GCD were carried out to assess the performance of the prepared ADC materials as electrodes for supercapacitor uses (Figure 6). The quasi-rectangular CV curves as well as the isosceles shape of the charge/discharge profile of the device indicated good capacitive performance of the ADC materials. The specific capacitances of these ADC electrode materials from the CV test and the GCD test matched each other, which decreased with the increase of the scan rate in the CV test and with the increase of current density in the GCD test.

The Raw-ADC showed relatively poor electrochemical performance as supercapacitor electrode material. It showed a specific capacitance of 24.4 F/g in the CV test at a scan rate of 5 mV/s and a specific capacitance of 16.6 F/g in the GCD test at a current density of 0.5 A/g. All the activation and doping processing resulted in ADCs with larger specific capacitances, but the N-doping and Fe-doping processing led to less improvement than that of the activation. Compared to the Raw-ADC, the Fe-ADC and N-ADC exhibited 2.3-fold and 3.9-fold specific capacitances, respectively, while the H_2SO_4-ADC, H_3PO_4-ADC, and HNO_3-ADC demonstrated 7.1-fold, 8.1-fold, and 10.9-fold specific capacitances, respectively. Particularly, the KOH-ADC presented the greatest improvement in specific capacitance, which reached 234.5 F/g during the GCD test at a current density of 0.5 A/g, i.e., 14.1-fold specific capacitance with respect to that of Raw-ADC. This result is better than the reported electrochemical performance of related carbon electrode materials from hydrothermal liquefaction processing of other biomass such as corn stover [12], pine wood [16], and larch wood [17]. It is noteworthy that the KOH-ADC exhibited 50% more specific capacitance than that of the carbon electrode materials from pyrolysis of the same microalgae [15].

To further characterize the electrochemical behavior of the ADC electrode materials, electrochemical impedance spectroscopy (EIS) was performed (Figure 7). The diameter of the semi-circle of the Nyquist plot in the high frequency region can be ascribed to electrolyte resistance, while the intercept of the Nyquist plot at the axis of the real part of complex impedance in the high frequency region is associated with the electrode resistance, which together make up the internal resistance of the electrode [27]. The angle between the straight line of the Nyquist plot and the axis of the real part of complex impedance indicates capacitive behavior of the electrode material. According to the Nyquist plots of all the ADC electrode materials, KOH-ADC showed the least electrolyte resistance, the second least electrode resistance, and the best capacitive behavior among all the ADC electrode materials, which is consistent with its best electrochemical performance.

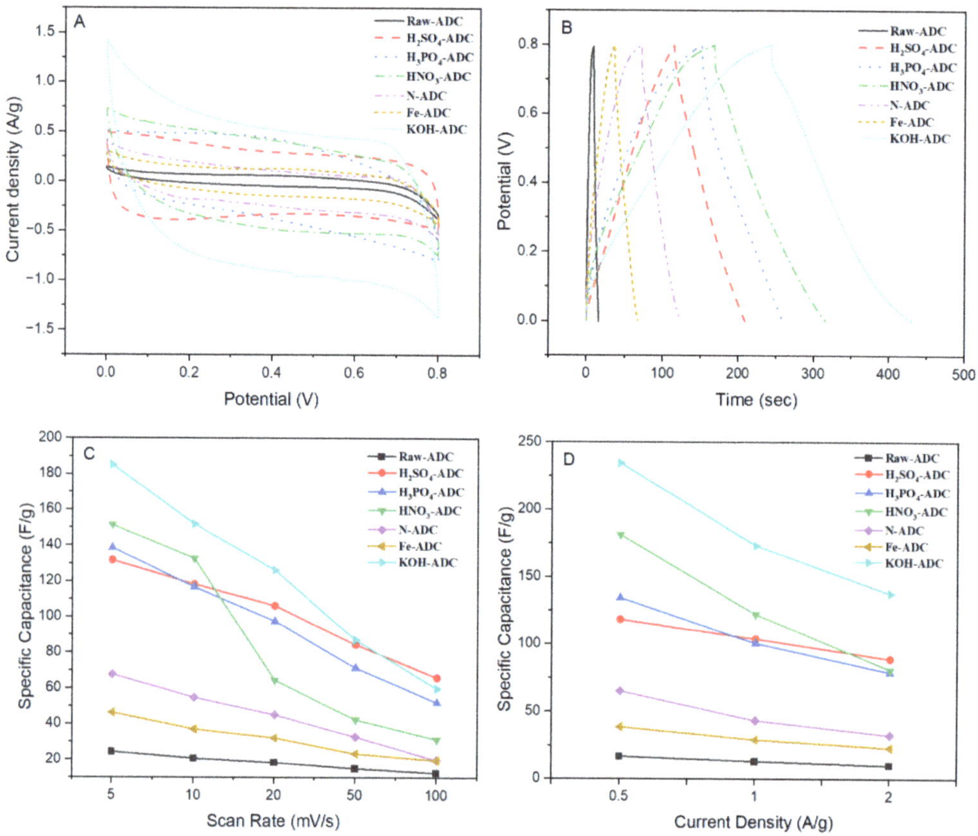

Figure 6. CV curves at scan rate of 5 mV/s (**A**), GCD profiles at current density of 0.5 A/g (**B**), specific capacitances from CV at various scan rates (**C**), and specific capacitances from GCD at various current densities (**D**) of all the ADC electrode materials.

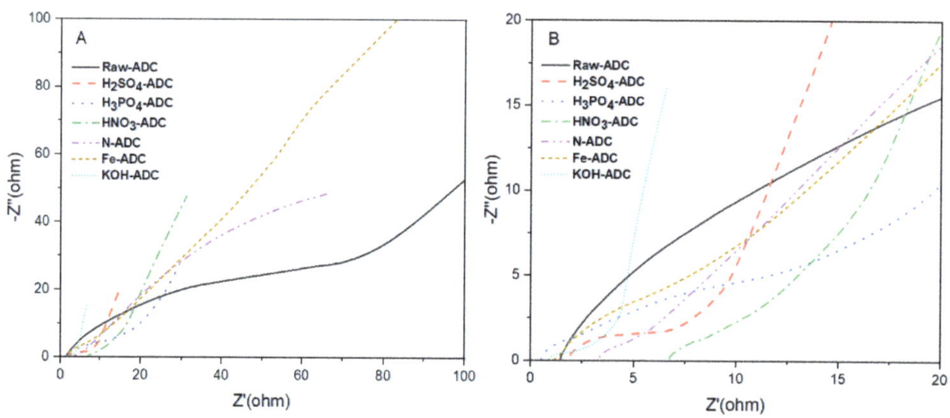

Figure 7. Nyquist plots of all the ADC electrode materials (**A**) and their views in a high frequency region (**B**).

3.4. Discussion

The electrochemical performance of all the ADC electrode materials can be correlated with their structure including specific surface area, pore structure, carbon structure, and surface properties. The final electrochemical performance of ADC electrode materials is a combination of all these factors.

KOH-ADC demonstrated the best electrochemical performance, which can be first attributed to its well-developed pore structure and consequently large specific surface area. The presence of the largest specific surface area (1418 m^2/g) and the biggest micropore volume (0.0587 cm^3/g) may have created the largest active sites for electrochemical activity to take place. In the meantime, the biggest mesopore volume has not only contributed to the specific surface area, but also created more accessible channels for the transport of ions to those micropore sites [28]. Overall, KOH-ADC electrode materials possessed the greatest capability for charge transport paths and storage sites.

Furthermore, KOH-ADC had the smallest average pore size (~2.2 nm) among all the ADC electrode materials, which is closest to the average $SO_4{}^{2-}$ electrolytic ionic radius (~0.258 ± 0.004 nm) [29]. Research has shown that a smaller average pore size allows for maximal ion interaction with pore walls, resulting in an increase in the EDL charge capacity without impeding the electro- adsorption and desorption kinetics [30–32].

The degree of ordered carbon structures in electrode materials influences the final supercapacitor performance. A relatively high degree of I_G/I_D implies a high level of graphitization in the electrode material, which can improve its electrical conductivity by reducing the charge transfer resistance to electrolyte ions [33], whereas a high level of a defective phase (low I_G/I_D) results in a high degree of electrolyte wetting of the electrode surface, which can lead to an enhancement in the electrochemical activity. The final electrochemical performance of carbon electrode materials for supercapacitors is a composite effect of both defective and graphitic phases that are present in the electrode material. From the Raman spectroscopy, KOH-ADC possessed a moderate I_G/I_D ratio (1.02) and therefore benefitted from the composite effect of both wetting and graphitization, resulting in the best electrochemical performance.

Additionally, KOH-ADC exhibited the largest atomic percentage of pyrrolic-N (85.13% from the XPS analysis), as well as the largest amount of O-containing functional groups (Table 2). It is established that pyrrolic-N improves electrochemical performance by enhancing the surface wettability properties of the carbon electrode as well as improving its pseudocapacitance [34]. Pyrrolic-N sites are located at the edge of carbon structure, and they consist of five-member rings that are bonded to the adjacent carbon atoms from the phenolic/carbonyl groups. The pyrrolic-N could contribute two π electrons to the aromatic system. The high proportion of pyrrolic-N on the surface of KOH-ADC may improve electrolyte wetting and facilitate the diffusion of electrolyte ions through the pores of the electrode material, as well as induce pseudocapacitive effects. The surface O-containing functional groups of KOH-ADC may further improve the electrolyte wetting properties. Overall, the well-developed porous structure and surface properties of KOH-ADC, in turn, resulted in a very low electrode resistance, rapid electrolyte diffusion, and good accessibility of the electrolyte ions to the electrode as evidenced by the Nyquist plot. The best specific capacitance 234.5 F/g among all the ADC electrode materials was thus acquired from KOH-ADC.

For comparison, the Raw-ADC performs poorly as an electrode material for supercapacitors (specific capacitance, 16.6 F/g) due to its poor porous structure, i.e., the largest average pore size of 33.51 nm, no micropores, and the lowest specific surface area. Fe-doping and N-doping did not significantly improve the electrochemical performance as anticipated, which could be contributed to the not-well-developed porous structure as evidenced by the lowest total pore volumes among all the prepared ADC electrode materials. Furthermore, their not-so-good electrochemical performance could also be ascribed to the relatively low amount of Fe atoms (Table 1) in the case of Fe-ADC and the 100% pyridinic-N structure in the case of N-ADC.

3.5. Practical Evaluation

The cycle stability test of a supercapacitor device is of prime importance as it provides the ground information for the practical use of the device. An outstanding capacitance retention of about 97% was attained for the high-performance KOH-ADC electrode material after 10,000 cycles at a current density of 10 A/g (Figure 8). The high capacitance retention of this electrode material could be attributed to the superior structural advantages of KOH-ADC, i.e., the large mesopore volume that created easily accessible channels to the micropores as well as its ultra-large specific surface area and abundant open pores on the surface for EDL formation. In addition, the device showed a stable coulombic efficiency (~55%) for stable long-term operation. The KOH-ADC electrode material also possessed an energy density of 5.1 W h/kg and a power density of 97.6 W/kg, indicating an overall good electrode material for supercapacitor use.

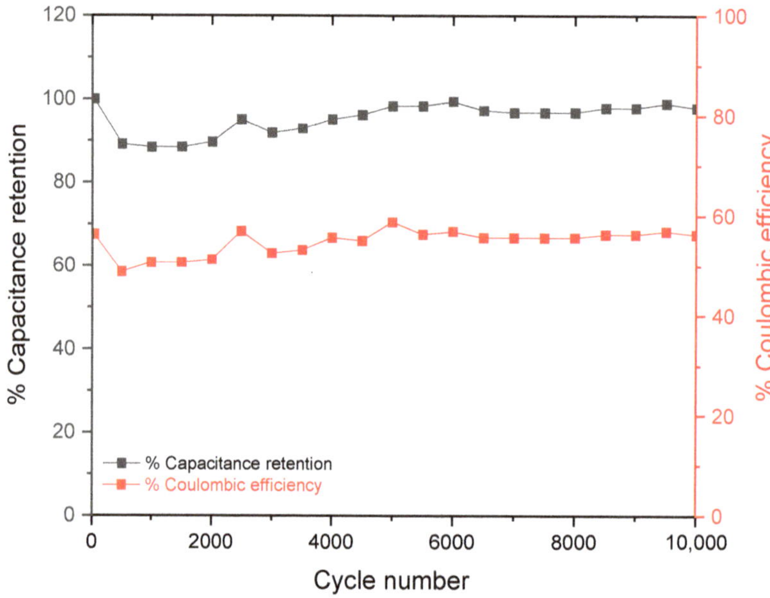

Figure 8. Cycle stability and coulomb efficiency of KOH-ADC electrode material up to 10,000 cycles of charging-discharging at current density of 10 A/g.

4. Conclusions

In this research, we demonstrated a successful attempt to acquire sustainable carbon electrode materials for supercapacitor use from the carbon byproduct after hydrothermal liquefaction processing of renewable algal biomass. With the intention to enhance the electrochemical performance of the obtained algae-derived carbon (ADC), we performed carbon activation with H_2SO_4, H_3PO_4, HNO_3, and KOH according to their different activation mechanisms, as well as N-doping with melamine and Fe-doping with iron (III) chloride. Compared to the original Raw-ADC, all the activation and doping processing resulted in ADCs with greater specific capacitances, but the N-doping and Fe-doping processing led to less improvement than that of the activation. Among all the activated and doped ADC electrode materials, KOH-ADC showed the best electrochemical performance and its specific capacitance reached 234.5 F/g through the galvanostatic charge-discharge (GCD) test at a current density of 0.5 A/g, which is a 14.1-fold increase with respect to that of the Raw-ADC. The electrochemical performance of KOH-ADC is better than that of the reported carbon electrode materials from the hydrothermal liquefaction processing of other biomass. This excellent electrochemical performance was supported by the least electrolyte resistance, the second least electrode resistance, and the best capacitive behavior among all the ADC electrode materials from the results of electrochemical impedance spectroscopy (EIS) test. The best electrochemical performance of the KOH-ADC can be attributed first to its well-developed pore structure with the smallest average pore size (2.2 nm), the biggest micropore volume (0.0587 cm^3/g), and the largest specific surface area (1418 m^2/g). Furthermore, the moderate I_G/I_D ratio of KOH-ADC warranted both good electrolyte wetting and electrical conductivity, while the largest atomic percentage of pyrrolic-N (88.13% from XPS) and the largest amount of O-containing functional groups in KOH-ADC also improved electrolyte wetting and diffusion of the electrolyte ions and facilitated pseudocapacitive effect. KOH-ADC further showed an outstanding capacitance retention (97%) after 10,000 cycles at a high current density of 10 A/g and a stable coulombic efficiency (~55%) for stable long-term operation. The KOH-ADC electrode material possessed an energy density of 5.1 W h/kg and a power density of 97.6 W/kg. This research points out a promising direction to expand use and add value to the carbon byproduct from the hydrothermal liquefaction of algae. In the meantime, this carbon byproduct is a sustainable carbon source, and its use will facilitate the sustainable development of supercapacitor electrode materials.

Supplementary Materials: The following supporting information can be downloaded at: https://www.mdpi.com/article/10.3390/c10020051/s1, Figure S1: High resolution N_{1s} XPS spectra of the ADCs with deconvoluted peaks of oxidized N, quaternary N, pyrrolic N, and pyridinic N: (A) Raw-ADC; (B) H_2SO_4-ADC; (C) H_3PO_4-ADC; (D) HNO_3-ADC; (E) N-ADC; (F) Fe-ADC; (G) KOH-ADC.

Author Contributions: Conceptualization: L.Z. and A.S.; data curation: K.A. and A.M.; formal analysis: K.A. and M.F.H.; funding acquisition and resources: A.S. and L.Z.; investigation and validation: K.A., A.M., M.F.H. and P.A.; methodology: K.A., A.M., M.F.H. and L.Z.; project administration and supervision: L.Z.; writing —original draft: K.A.; writing—review and editing: L.Z. All authors have read and agreed to the published version of the manuscript.

Funding: This work was partially supported by the United States Department of Agriculture/National Institute of Food and Agriculture (USDA/NIFA) Evans-Allen Fund (NC. X331-5-21-130-1). The authors also thank the support from the Joint School of Nanoscience and Nanoengineering of North Carolina A&T State University, a member of the Southeastern Nanotechnology Infrastructure Corridor (SENIC) and the National Nanotechnology Coordinated Infrastructure (NNCI), which is financially supported by the National Science Foundation (ECCS-2025462).

Data Availability Statement: The original contributions presented in the study are included in the article/Supplementary Material, further inquiries can be directed to the corresponding author/s.

Conflicts of Interest: The authors declare no conflicts of interest.

References

1. Bilgen, S.; Kaygusuz, K.; Sari, A. Renewable energy for a clean and Sustainable future. *Energy Sources* **2004**, *26*, 1119–1129. [CrossRef]
2. Olabi, A.G.; Abbas, Q.; Al Makky, A.; Abdelkareem, M.A. Supercapacitors as next generation energy storage devices: Properties and applications. *Energy* **2022**, *248*, 123617. [CrossRef]
3. Yu, Z.; Tetard, L.; Zhai, L.; Thomas, J. Supercapacitor electrode materials: Nanostructures from 0 to 3 dimensions. *Energy Environ. Sci.* **2015**, *8*, 702–730. [CrossRef]
4. Abruna, H.D.; Yasuyuki, K.; Henderson, J.C. Batteries and electrochemical capacitors. *Phys. Today* **2008**, *61*, 43–47. [CrossRef]
5. Fic, K.; Platek, A.; Piwek, J.; Frackowiak, E. Sustainable materials for electrochemical capacitors. *Mater. Today* **2018**, *21*, 437–454. [CrossRef]
6. Gao, Z.; Zhang, Y.; Song, N.; Li, X. Biomass-derived renewable carbon materials for electrochemical energy storage. *Mater. Res. Lett.* **2017**, *5*, 69–88. [CrossRef]
7. Akhtar, J.; Amin, N.A.S. A review on process conditions for optimum bio-oil yield in hydrothermal liquefaction of biomass. *Renew. Sustain. Energy Rev.* **2011**, *15*, 1615–1624. [CrossRef]
8. Ponnusamy, V.K.; Nagappan, S.; Bhosale, R.R.; Lay, C.-H.; Nguyen, D.D.; Pugazhendhi, A.; Chang, S.W.; Kumar, G. Review on sustainable production of biochar through hydrothermal liquefaction: Physico-chemical properties and applications. *Bioresour. Technol.* **2020**, *310*, 123414. [CrossRef]

9. He, S.; Barati, B.; Hu, X.; Wang, S. Carbon migration of microalgae from cultivation towards biofuel production by hydrothermal technology: A review. *Fuel Process. Technol.* **2023**, *240*, 107563. [CrossRef]
10. Leng, L.; Yuan, X.; Zeng, G.; Shao, J.; Chen, X.; Wu, Z.; Wang, H.; Peng, X. Surface characterization of rice husk bio-char produced by liquefaction and application for cationic dye (Malachite green) adsorption. *Fuel* **2015**, *155*, 77–85. [CrossRef]
11. Salehi, B.; Wang, L. Critical Review on Nanomaterials for Enhancing Bioconversion and Bioremediation of Agricultural Wastes and Wastewater. *Energies* **2022**, *15*, 5387. [CrossRef]
12. Shell, K.M.; Rodene, D.D.; Amar, V.; Thakkar, A.; Maddipudi, B.; Kumar, S.; Shende, R.; Gupta, R.B. Supercapacitor performance of corn stover-derived biocarbon produced from the solid co-products of a hydrothermal liquefaction process. *Bioresour. Technol. Rep.* **2021**, *13*, 100625. [CrossRef]
13. Gao, Y.; Sun, R.; Li, A.; Ji, G. In-situ self-activation strategy toward highly porous biochar for supercapacitors: Direct carbonization of marine algae. *J. Electroanal. Chem.* **2021**, *882*, 114986. [CrossRef]
14. Kandasamy, S.; Bhuvanendran, N.; Narayanan, M.; He, Z. Chapter 13—Thermochemical conversion of algal biomass. In *Handbook of Algal Biofuels*; Elsevier Inc.: Amsterdam, The Netherlands, 2022; pp. 281–302.
15. Yardim, Y.; Saka, C. Oxygen and nitrogen-doped carbon particles derived from pyrolysis of Chlorella vulgaris and Spirulina platensis microalgae as an efficient electrode material for supercapacitor application. *Full. Nanotub. Carbon Nanostructures* **2023**, *31*, 713–723. [CrossRef]
16. Amar, V.S.; Houck, J.D.; Shende, R.V. Catalytic HTL-derived biochar and sol-gel synthesized (Mn, Ti)-oxides for asymmetric supercapacitors. *Int. J. Energy Res.* **2020**, *44*, 12546–12558. [CrossRef]
17. Hwang, H.; Lee, J.-H.; Ahmed, M.A.; Choi, J.W. Evaluation of pyrochar and hydrochar derived activated carbons for biosorbent and supercapacitor materials. *J. Environ. Manag.* **2021**, *298*, 113436. [CrossRef]
18. Hasan, M.F.; Mantripragada, S.; Gbewonyo, S.; Xiu, S.; Shahbazi, A.; Zhang, L. Carbon nanofibrous electrode material from electrospinning of chlorella (microalgae) with polyacrylonitrile for practical high-performance supercapacitor. *Int. J. Energy Res.* **2022**, *46*, 22867–22882. [CrossRef]
19. Hasan, M.F.; Asare, K.; Mantripragada, S.; Charles, V.; Shahbazi, A.; Zhang, L. Meso-microporous carbon nanofibrous aerogel electrode material with fluorine-treated wood biochard for high-performance supercapacitor. *Gels* **2024**, *10*, 82. [CrossRef]
20. Nzediegwu, C.; Naeth, M.A.; Chang, S.X. Feedstock type drives surface property, demineralization and element leaching of nitric acid-activated biochars more than pyrolysis temperature. *Bioresour. Technol.* **2022**, *344 Pt B*, 126316. [CrossRef]
21. Yorgun, S.; Yıldız, D. Preparation and characterization of activated carbons from Paulownia wood by chemical activation with H_3PO_4. *J. Taiwan Inst. Chem. Eng.* **2015**, *53*, 122–131. [CrossRef]
22. Wang, G.; Zhang, J.; Kuang, S.; Zhou, J.; Xing, W.; Zhuo, S. Nitrogen-doped hierarchical porous carbon as an efficient electrode material for supercapacitors. *Eletrochimica Acta* **2015**, *153*, 273–279. [CrossRef]
23. Li, M.; Xue, J. Integrated synthesis of nitrogen-doped mesoporous carbon from melamine resins with superior preformance in supercapacitors. *J. Phys. Chem. C* **2014**, *118*, 2507–2517. [CrossRef]
24. Wu, D.; He, T.; Albashir, A.I.M.; Tan, Y.; Ran, F. Iron-doped carbon electrode materials derived from polyethersulfone. *J. Energy Storages* **2021**, *33*, 102099. [CrossRef]
25. Wang, J.; Kaskel, S. KOH activation of carbon-based materials for energy storage. *J. Mater. Chem.* **2012**, *22*, 23710–23725. [CrossRef]
26. Zhu, L.; Zhao, N.; Tong, L.; Lv, Y. Structural and adsorption characteristics of potassium carbonate activated biochar. *RSC Adv.* **2018**, *8*, 21012–21019. [CrossRef]
27. Mei, B.-A.; Munteshari, O.; Lau, J.; Dunn, B.; Pilon, L. Physical Interpretations of Nyquist Plots for EDLC Electrodes and Devices. *J. Phys. Chem. C* **2018**, *122*, 194–206. [CrossRef]
28. Asare, K.; Hasan, F.; Shahbazi, A.; Zhang, L. A comparative study of porous and hollow carbon nanofibrous structures from electrospinning for supercapacitor electrode material development. *Surf. Interfaces* **2021**, *26*, 101386. [CrossRef]
29. Jenkins, H.D.B.; Thakur, K.P. Reappraisal of Thermochemical Radii for Complex Ions. *J. Chem. Educ.* **1979**, *56*, 576–577. [CrossRef]
30. Vicentini, R.; Nunes, W.G.; Costa, L.H.; Pascon, A.; da Silva, L.M.; Baldan, M.; Zanin, H.G. Environmentally friendly functionalization of porous carbon electrodes for aqueous based electrochemical capacitors. *IEEE Trans. Nanotechnol.* **2018**, *18*, 73–82. [CrossRef]
31. Eliad, L.; Pollak, E.; Levy, N.; Salitra, G.; Soffer, A.; Aurbach, D. Assessing optimal pore-to-ion size relations in the design of porous poly(vinylidene chloride) carbons for EDL capacitors. *Appl. Phys. A Mater. Sci. Process.* **2006**, *82*, 607–613. [CrossRef]
32. Largeot, C.; Portet, C.; Chmioia, J.; Taberna, P.-L.; Gogotsi, Y.; Simon, P. Relation between the Ion Size and Pore Size for an Electric Double-Layer Capacitor. *J. Am. Chem. Soc.* **2008**, *130*, 2730–2731. [CrossRef] [PubMed]
33. Ranaweera, C.K.; Kahol, P.K.; Ghimire, M.; Mishra, S.R.; Gupta, R.K. Orange-Peel-Derived Carbon: Designing Sustainable and High-Performance Supercapacitor Electrodes. *C* **2017**, *3*, 25. [CrossRef]
34. Hulicova-Jurcakova, D.; Seredych, M.; Lu, G.Q.; Bandosz, T.J. Combined Effect of Nitrogen- and Oxygen-Containing Functional Groups of Microporous Activated Carbon on its Electrochemical Performance in Supercapacitors. *Adv. Funct. Mater.* **2009**, *19*, 438–447. [CrossRef]

MDPI AG
Grosspeteranlage 5
4052 Basel
Switzerland
Tel.: +41 61 683 77 34

C Editorial Office
E-mail: carbon@mdpi.com
www.mdpi.com/journal/carbon

www.ingramcontent.com/pod-product-compliance
Lightning Source LLC
Chambersburg PA
CBHW062053220926
51850CB00010B/2756
* 9 7 8 3 7 2 5 8 4 6 6 3 4 *